I0199271

Crystal Healing for Your Chakras:

The TRUE Call of Nature

A Beginner's Introduction to the World of Healing Crystals

Blue Agate

Fire Agate

Crystal Healing for Your Chakras:
The TRUE Call of Nature
A Beginner's Introduction to the World of Healing Crystals

Liam J. Adair

Lemurian Quartz

Wholesome Healing Consultants, LLC

2020

This and prior editions published in 2019, 2020, & 2021 by:
Wholesome Healing Consultants, LLC
Houston, TX
https://www.whliam.com

Note to the Reader: *This book is intended as an informational guide. The remedies, approaches, and techniques described herein are meant to supplement, and not meant to be a substitute for, professional medical care or treatment. They should not be used to treat a serious ailment without prior consultation with a qualified health care professional.*

First Printing: December 2019

10 9 8 7 6 5 4 3 2 1

A Library of Congress CIP (Cataloging-in-Publication) Number and copy of this book is available from the Library of Congress. Control Number: 2019920971

ISBN Hardcover: 978-0-578-62098-5
ISBN KDP: 978-1-671-17838-0 & 979-8-792-89260-6

Photography:
Liam J. Adair (Author)
Tiffany Smith (Maati Ra) and Samuel Leger (Sam Night) of Blue Treasure Photography; BlueTreasurePhotography.com: Cover, 10, 52, 106
iStock, Shutterstock, & Pixabay

Ordering Information:

Wholesale discounts are available on quantity purchases by corporations, associations, educators, and others.

U.S. trade bookstores and wholesalers: Please contact Amazon and hardcover printing agencies for direct shipping and printing. You may also contact Wholesome Healing Consultants LLC for more information: Tel: (214) 930-6747; https://whliam.com

Dedication

I dedicate this work of knowledge to my amazing mother,
Shirley Jean Adair.
Thank you mom for your undying and
unyielding love and support.

Contents

THANK
YOU

Acknowledgements

I would like to thank my mother, my coaches, instructors, friends, family and clients without whose help this book would not have been completed. Moni, you are always a great partner with which to brainstorm.

Lee, thank you for your insight and your keen eyes for editing. Special thanks to Blue Treasure Photography for your beautiful works of art.

The community, strengthening our 60lb Rose Quartz, receives loving energy in return.

Chalcopyrite (Peacock Ore)

Pyrite (Fool's Gold)

Preface

This book was written to further enlighten those who are new to and those who are experienced in working with Crystal and Stones. My mission is to provide useful information that fosters an open mind and heart towards healing your Body, Mind & Spirit, using the Gifts of Nature. Neither a person's way of life nor their ritualistic practice restricts them from accessing and applying the knowledge from within.

Agate

It serves as an introduction into the energy realm of gemstones. You will learn of a few of the methods in practice used to detect energy and the best form of meditation for Crystal Healing Therapy. Allow your intuition to drive your curiosity. Allow that curiosity to drive you to formulate questions. Embrace the sense of comfort, as well as the moments of discomfort brought to the surface; for through great discomfort spurs exponential growth. Allow your "misunderstandings" to spark your curiosity. Allow that curiosity to convey questions you may have never thought to ask or may have been too fearful to face. Seek those answers with a "child-like" curiosity; be willing to learn.

Let us be Present, Let us be of Light, Let us be of Love

Introduction

Alignment, appropriate relative position of agreement. Balance, an even distribution, evenly supporting all parts of the whole. When we consider balance of the chakras, we are aiming to align. One chakra cannot maintain balance if the others are not in alignment. All chakras are connected just as we Beings are connected to one another and the World in which we live. One chakra channel's condition can directly or indirectly affect another channel's energy flow. On that same notion, the health of our Being can directly or indirectly affect our families, careers, and the environment.

This understanding drives the importance for bringing balance to ourselves; starting with cleansing and aligning our chakras and allowing the positivity of that balance to flow into all other areas of our lives.

ENERGY POINTS OF THE SUBTLE BODY

CROWN CHAKRA
SAHASRARA
COSMIC ENERGY
BLOCKED BY ATTACHMENT
SILENCE

THIRD EYE CHAKRA
AJNA
INTUITION
BLOCKED BY ILLUSION
OM

THROAT CHAKRA
VISHUDDHA
TRUTH
BLOCKED BY LIES
HAM

HEART CHAKRA
ANAHATA
LOVE
BLOCKED BY GRIEF
YAM

SOLAR PLEXUS CHAKRA
MANIPURA
WILLPOWER
BLOCKED BY SHAME
RAM

SACRAL CHAKRA
SVADHISTHANA
PLEASURE
BLOCKED BY GUILT
VAM

ROOT CHAKRA
MULADHARA
SURVIVAL
BLOCKED BY FEAR
LAM

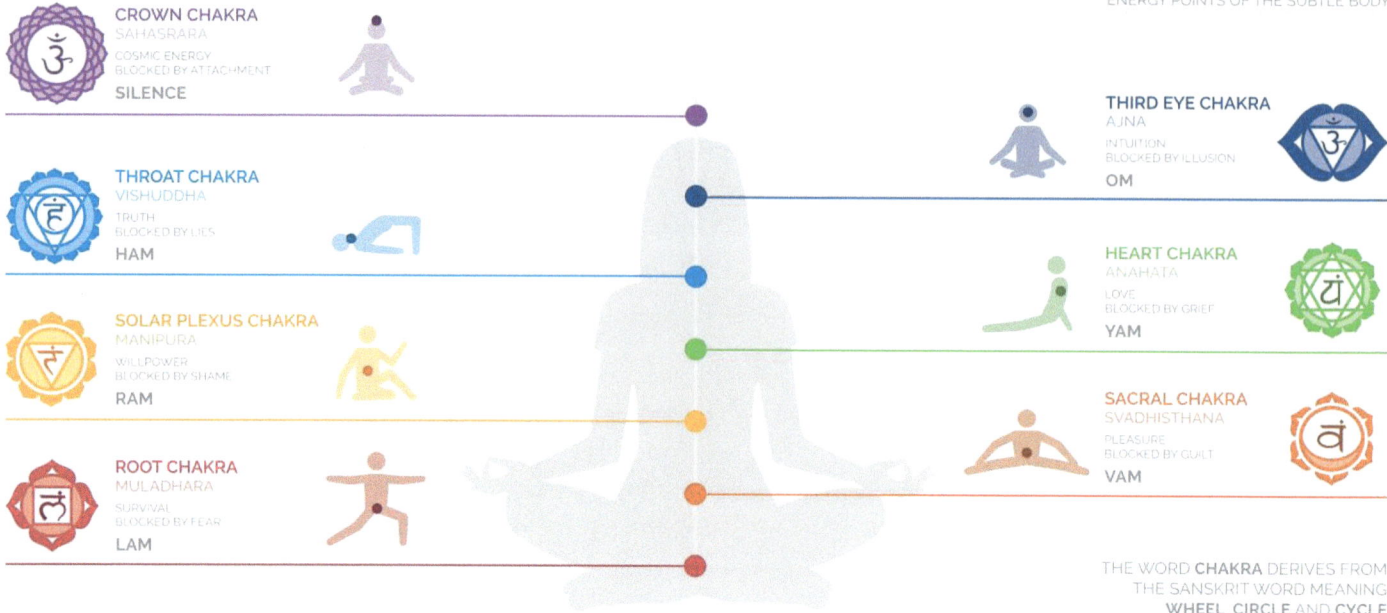

THE WORD **CHAKRA** DERIVES FROM THE SANSKRIT WORD MEANING **WHEEL**, **CIRCLE** AND **CYCLE**

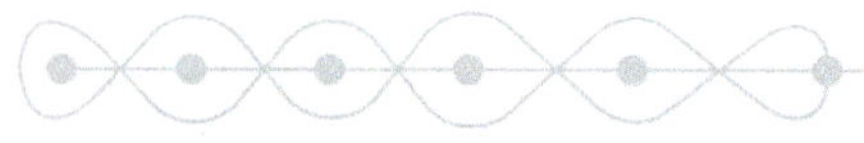

THE CHAKRAS
ENERGY HEALING

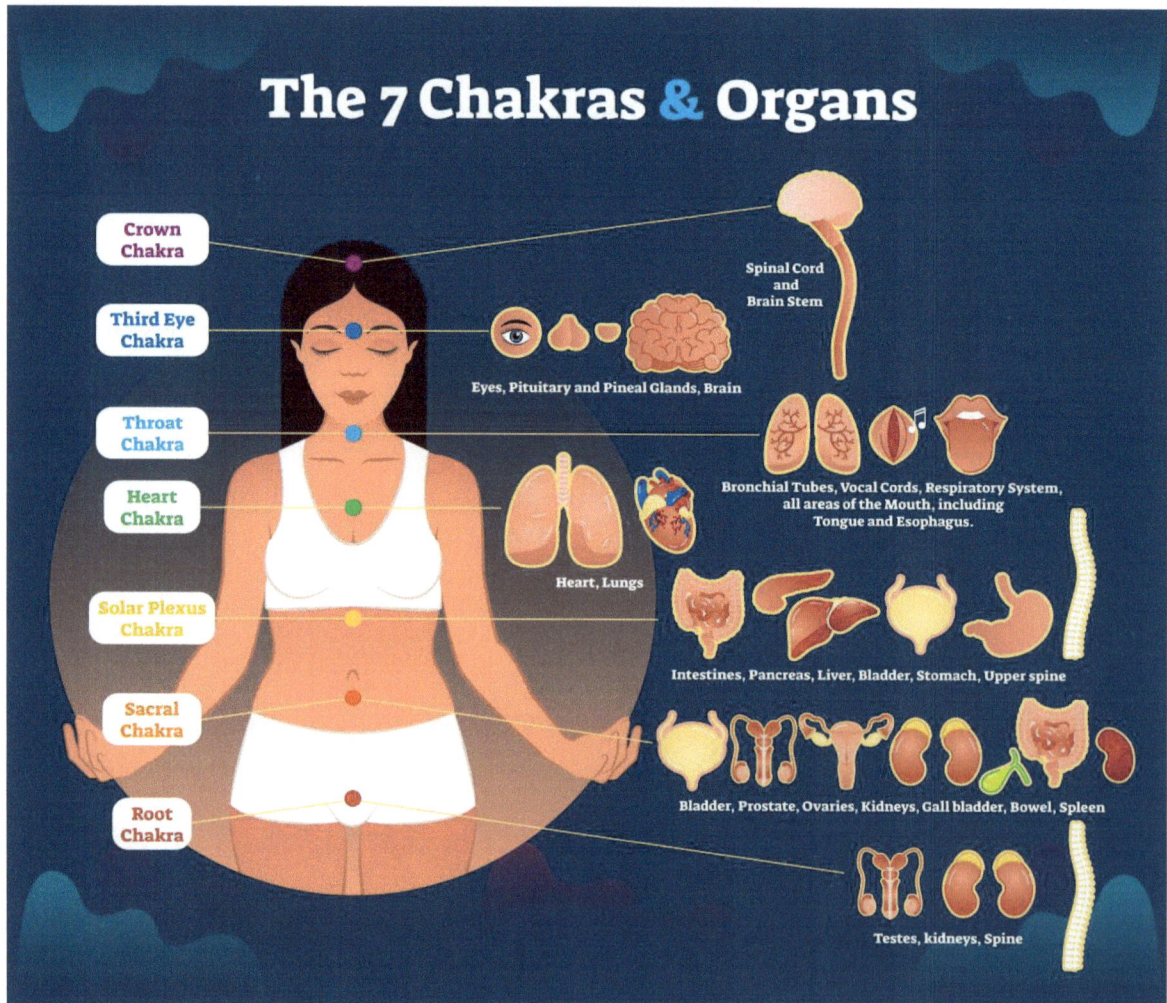

The 7 Chakras & Organs

- **Crown Chakra** — Spinal Cord and Brain Stem
- **Third Eye Chakra** — Eyes, Pituitary and Pineal Glands, Brain
- **Throat Chakra** — Bronchial Tubes, Vocal Cords, Respiratory System, all areas of the Mouth, including Tongue and Esophagus.
- **Heart Chakra** — Heart, Lungs
- **Solar Plexus Chakra** — Intestines, Pancreas, Liver, Bladder, Stomach, Upper spine
- **Sacral Chakra** — Bladder, Prostate, Ovaries, Kidneys, Gall bladder, Bowel, Spleen
- **Root Chakra** — Testes, kidneys, Spine

Chapter 1: The Bases of Chakra Explored

To understand just how to make chakra alignment a reality, we must first discuss the chakras. What are they? What is their purpose in relation to the body and the Universe?

The chakras are the channels or stores for energy, the spirit, power, life source, Qi (pronounced Chi, as said in China), Mana (as said by Kahunas of Hawaii), Prana (as said of the Hindu), Ki (as said in Japan) or the label that best represents your beliefs. In Sanskrit, the word chakra translates to "wheel", "circle", or "cycle".

These stores, or powerhouses, draw energy from the universe into the human and animal body. This life source feeds and brings balance to the body's energy level, which in turn feeds the organs and vital systems to produce a healthy functioning Being. At the chakra points, the energy flow rotates, like a wheel. The outer portion of the wheel of energy is situated about four to six inches from the body. Each wheel spins either clockwise or counter clock wise. The chakras appear to revolve in the opposite directions to the chakras above and below it.[1]

Ways of Detecting Chakras & Body Energy Fields

Using your hands, you are able to feel your own energy flow and the flow of another's. If you hover your hand over one of your chakra points, you will feel a flow of heat. As you practice recognizing it, you will be able to easily detect, what has been described as, the soft rise and fall of a heart beat; similar to a warm slowly-pulsing sensation gently brushing against your palm.

In addition to hand techniques, there are a number of ways in which you can detect and in some cases even see the body's energy field and the flow of Qi. One such technique is by using the Gas Discharge Visualization (GVD) device. Another detection method is via the practice of Reiki Healing, which is an advanced healing technique using the hands to sense and emit healing energy. One may also use a pendulum or psychic and spiritual aura readings. This list is not all inclusive, but we have learned that detection can be accomplished with both technological devices and organic Beings.

Gas Discharge Visualization (GDV)

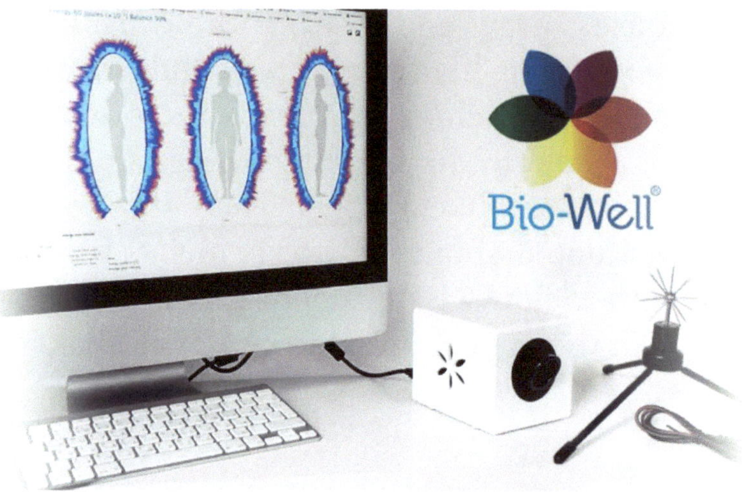

We have witnessed advances in energy detection in reference to the body. Other than by seeing Auras directly, these techniques helped to paint the picture of what we have intuitively known to exist since the dawn of man. Konstantin Korotkov, professor at St. Petersburg

Federal University of Informational Technologies Optics in Mechanics in Russia, developed, what he calls, GVC. Since its creation, it has been enhanced to a more precise electro-photonic imaging machine named the Gas Discharge Visualization (GDV) device. Korotkov's device is considered a breakthrough compared to the original Kirlian photography, due to its ability to present direct, real-time viewing of the human energy system. His team explains that this technology allows one to capture, by a video camera, the electro-photonic glow from plants, inanimate objects, individuals, stones, liquids and powders and then translate it into a computerized model. This provides a platform for the healer and clients to obtain a visual record of the imbalances affecting their whole "body–mind–spirit" well-being. Not specifically a medical device, the GDV can be used in varying fields; including sound therapy, genetics, Biophysics, Psychology, Crystal Healing Therapy, agriculture, and the list goes on. The latest instrument, Bio-Well GDV camera, is being produced in Hong Kong and carries European Union (EU) and FDA certifications.[2]

Usui Reiki Healing

Translation: Reiki

The history of the Reiki Touch Healing began in the India-Nepal region around 620 BCE (Before Common Era). As years passed, various energy healing techniques were formulated and taught. Around 46 or 49 CE (Common Era) Buddha and the great Physicians, one of which who was Jesus of Nazareth - having survived the crucifixion and returned to India - practiced and taught these physical and spiritual healing techniques. Teachers and students alike recorded these events and, by selection only, taught these secrets to those they felt were deserving of the knowledge.[3]

In the late 1800s, Mikao Usui, principal of the Doshisha University in Koyoto, Japan and Christian Minister, set off on what became a 10-year quest to learn the methods by which Jesus conducted his healings. Receiving no answers through Christian sources, he continued his pursuit by studying other ritualistic practices.[4]

Through research, he discovered An-shin Ritsu-mei, pronounced ŏhn-sheen (r)eet-sooh-meh-ēe, a state of consciousness, once reached, helps one understand their life's purpose and how to actualize it. He made it his mission to reach this state. He learned that one could reach An-shin Ritsu-mei by practicing Zazen meditation. Eventually making his way to a monastery in Japan, he began studying under a Zen teacher. After three years without success, Usui Sensei's teacher suggested he travel up the sacred mountain, Kuruma Yama; advising that in order to reach his goal, he must be willing to die. Misunderstanding the advice, Usui Sensei willingly traveled up the mountain 'prepared to die'. On the mountain, he fasted and meditated for twenty days and nights. On the twenty first day, at the stroke of midnight,

Human Energy Scheme

> *"he stood up and a powerful light suddenly entered his mind through the top of his head and he felt as if he had been struck by lightning; this caused him to fall down unconscious." –William Lee Rand, 1991*

After waking, he was so excited that he began running down the mountain as quickly as he could. He was determined to share his joyous discovery with his Zen instructor. Not paying attention, he tripped and stumped his toe. Instinctually, he grabbed his hurting and throbbing toe and realized, after a few seconds, his pain was completely gone. This was accomplished due to Healing Energy flowing from his hands into his foot. In this moment of realization, he understood his life's purpose was to be a healer. He began structuring and perfecting the healing technique we identify as Usui Reiki.

Throughout the remainder of his life, he healed many and eventually passed the knowledge on to a chosen successor, Chujiro Hayashi.[5] [6] Hayashi's chosen successor, Hawayo Takata, was pivotal in Reiki Healing being established in the West. As a master of Usui Traditional Reiki, she trained hundreds of people, spreading Reiki Healing to all corners of the globe.

Anyone with the desire can learn to use Reiki Healing methods. Using the Reiki therapy techniques, one is able to detect areas on the body that are in need of healing and be effective conduits to deliver that healing. Using crystals and stones in combination with Reiki Touch will enhance and speed up the time it would normally take to address and relieve certain ailments.[7]

Onyx (Banded)

Pendulums

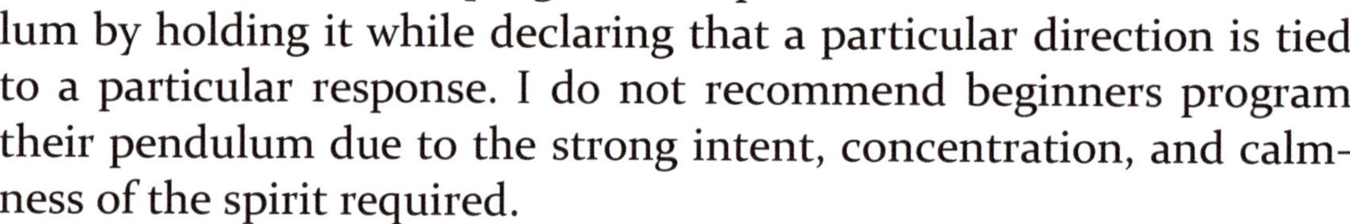

Pendulums are used quite often by alternative practitioners. Seeking knowledge of what is not readily available to the conscious mind can be interpreted by reading the sway of a pendulum. Simply stated, the pendulum is an object hanging from one end of a string, rope or chain. Prior to asking your questions, you need to determine the response direction or program your pendulum's responses. Pendulums can swing clockwise, counterclockwise, back and forth, side to side, diagonal and even vibrate. You can program the pendulum by holding it while declaring that a particular direction is tied to a particular response. I do not recommend beginners program their pendulum due to the strong intent, concentration, and calmness of the spirit required.

The easiest way for beginners to accomplish reading the energy driven responses is to sit comfortably in a chair with your feet flat on the ground. Using your dominant hand, hold your pendulum off to the side, about a foot from your body. Calm, focused and concentrating, ask the questions:

"Which response is yes?"

"Which response is no?"

After asking each question, one at a time, wait about 30 seconds for your pendulum to sway. You may ask a test question using your name or age to determine if your pendulum is attuned to you and responding accurately. If it is more comfortable for you, you may sit at a table, ensuring that your elbow is the only part of your body resting on it. This offers more stability for maintaining your arm's position during testing.[8]

At this time, you may now verify your chakra energy levels. To complete this task, one would intently hold their pendulum over the chakra channel in question. The healer, or client themselves would then formulate a question in their mind or speak it aloud. Again, it is very important to remain calm and focused while conducting question and answer sessions. Questions you may want to ask:

"Is this a healthy flow of chakra energy for my (specify the chakra)?"

"Is my (specify the chakra) flowing clockwise/counterclock-wise?"

Dowsing crystals (from left to right): Smoky Quartz, Rose Quartz, Thulite, Cavansite, Pyrite, Andalusite "Cross Stone", Quartz Crystal, & Peacock Ore

For Sahasrara (Crown) and Ajna (Third-eye) chakra, it helps to use a mirror. Another method by which to measure your chakra levels using the pendulum is done using a sheet of paper and a pen. Write your name at the top of the paper. This indicates that your intention is to focus on the chakra levels of your Being. Writing your name is not a requirement, it only serves to help focus your intent. Afterwards, draw a human figure to represent your body. Masterful artistry is not needed, a simple stick figure will work just fine. Next, draw circles along the chakra centers. Then, as you would do over your physical body, hover the pendulum over the circles you drew and ask questions to determine your current levels.

Cavansite

Mastering this tool helps you to easily and quickly determine which areas of your body needs to be addressed well before physical symptoms of 'dis-ease' manifest.

Psychic & Spiritual Aura Readings

Opal

The aura is a field of energy surrounding the body with a radius of up to sixteen or so feet. Visually, this energy field can be any one of or combination of the hues on the color spectrum. There are main or base colors and fluctuating colors. Your base colors, fundamentally represent your birth signature, personality traits, likely behavior, and can provide insight into your destiny potential. The fluctuating colors within your aura can reveal phenomena such as if you are suffering from a cold, if you are experiencing a panic attack and even when you are in a joyful mood.

Natural Red Ruby

Your aura is a complex field comprised of different levels. The photo identifies the major areas of your field. For the purposes of easily identifying the levels, I have used different colors. Remember, your aura field's base colors will be determined by your unique energy signature and the fluctuating colors in the outer fields are influenced by your current state of being.

Psychic and spiritual readings are intuition driven forms of energy detection in or around the body. Measuring chakra energy levels using your psychic or spiritual abilities allows one to accomplish readings at a distance. Oftentimes, we can visually observe someone and intuitively read their energy levels. There are many defined categories of psychic such as clairvoyant (to see), medium (to commune with), clairsentience (to feel), clairaudience (to hear), and channel (to receive messages). To be highly intuitive is a combination of two or more of these listed abilities. You do not have to be formally trained to possess these gifts.

Black Opal

Many young children are born with the ability to see auras. As adults, it is easy to mistake a child drawing a portrait of someone using a green crayon as a 'lack of skills'. I encourage you to inquire of their choice of color. You might be surprised to receive a response similar to "They look green to me." Assessing in the short term, a vibrant green may indicate the emotion of envy or an illness. As a base color, green indicates someone who is easy-going, peace loving, and a natural healer[9]. Developing and using this skill, one can create a healing opportunity for all that they encounter, regardless of the setting.

Chrysocolla

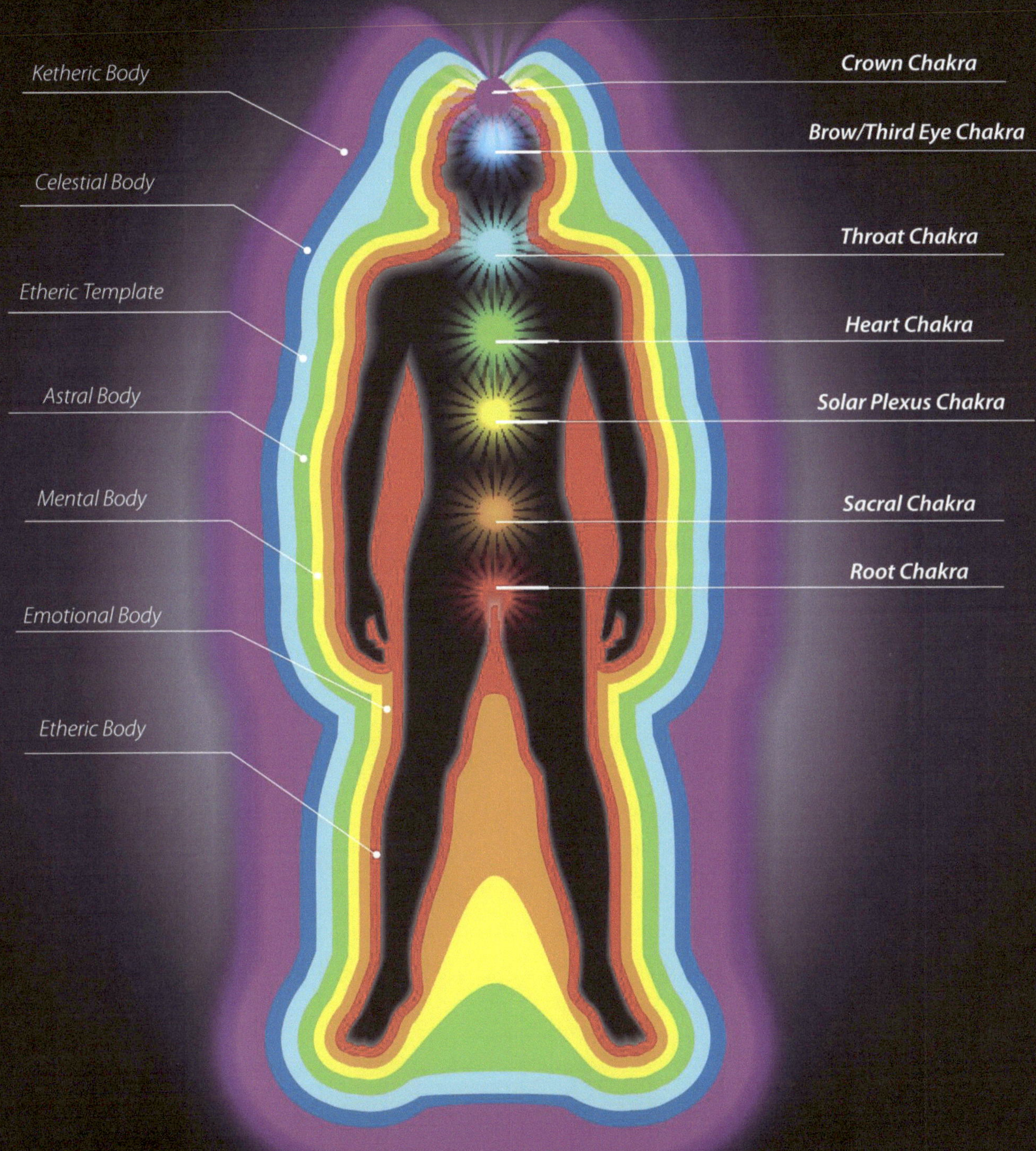
Ketheric Body
Celestial Body
Etheric Template
Astral Body
Mental Body
Emotional Body
Etheric Body
Crown Chakra
Brow/Third Eye Chakra
Throat Chakra
Heart Chakra
Solar Plexus Chakra
Sacral Chakra
Root Chakra

Chapter 2: The 7 Primary Chakras Defined

Defining the system of the chakras helps us to better understand how they work and affect the Being. There are seven major chakras and each chakra corresponds to a particular part of the physical body, emotional and mental body, and spiritual body. Generally, when one chakra is out of alignment, blocked, or weak, it is verified by a manifestation of dis-ease within its corresponding 'region'. Chakras cannot be fully blocked. As long as the Being has life, energy will always flow through the chakras. Even if limbs or physical parts have been removed, your morphogenetic field, the energy center which provides instructions for cells at the molecular level, helps to maintain the areas where your chakra centers are saturated. Once realigned and balanced, the present 'dis-ease' is relieved. It may be necessary to regularly repeat healing sessions. I have worked with clients who have formed lasting relationships with crystals and stones because maintenance calls for continued use of those particular gemstones. The ability of nature to heal the Body, Mind, and Spirit knows no bounds.

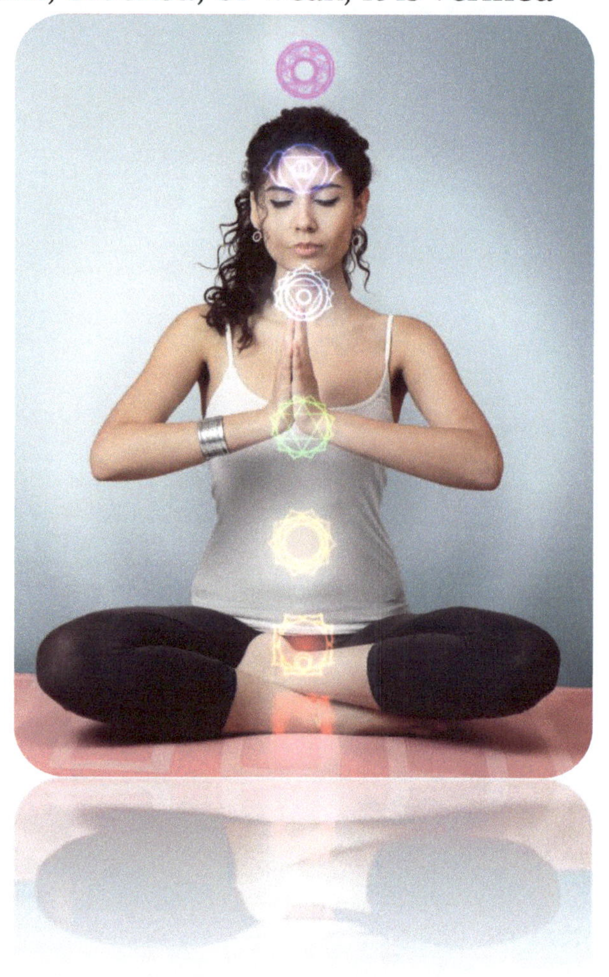

Let us dive into defining and identifying the chakras. The centers of the seven chakras run along the spine of the physical body. The centers also have a corresponding area on the anterior side of the body. For ease of understanding its position, when I define the chakras, I will identify the centers using positions from the body's frontal view.

Possessing a strong awareness of the chakras is the gateway to understanding how our life force interacts with our bodies and the Universe at large. Creation is stirred by our energies and with balancing and healing our inner selves, we are able to gain insight into the construction of existence. An amazing representation of Creation is characterized by the Tree of Life.

Tree of Life: Sephirot

The life force of our chakras interact with our entire Being in accordance with a Grand Design. Who or what the original architect is depends upon your beliefs and what aligns with your morals. Our thought patterns, behaviors and actions both influence and follow

our interpretation of the Design. There are many representations of the Grand Design and one such example is the Tree of Life.

The Tree of Life consists of ten Sephirot, or numbers, with 22 Paths between them. The Tree of Life is not the definitive representation of the complexities of this amazing Universe in which we exist; however, this is one of the best representations mankind can devise and comprehend to barely scratch the surface of what we believe we know of Creation.

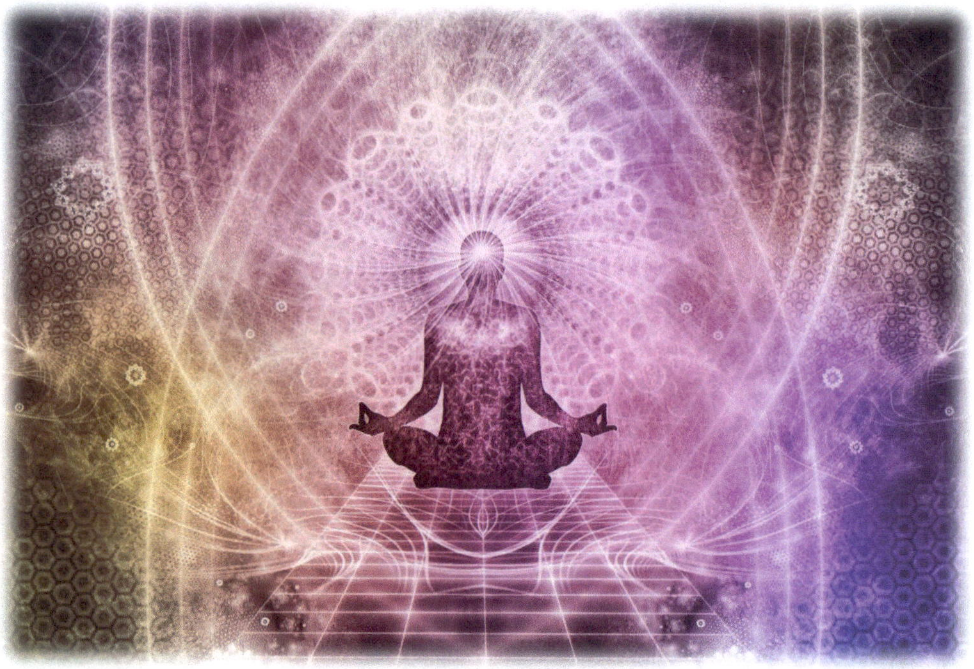

An understanding becoming more common is the belief that all creation, everything we see, feel, experience, and dream, formed from "No-thing". Absolute zero is absolutely nothing. For if there is to be the very first something, by comparison, that something had to have come from the very first nothing. A spark, to exist, to breathe, to blink, to experience, to manifest, is represented by the Tree of Life.

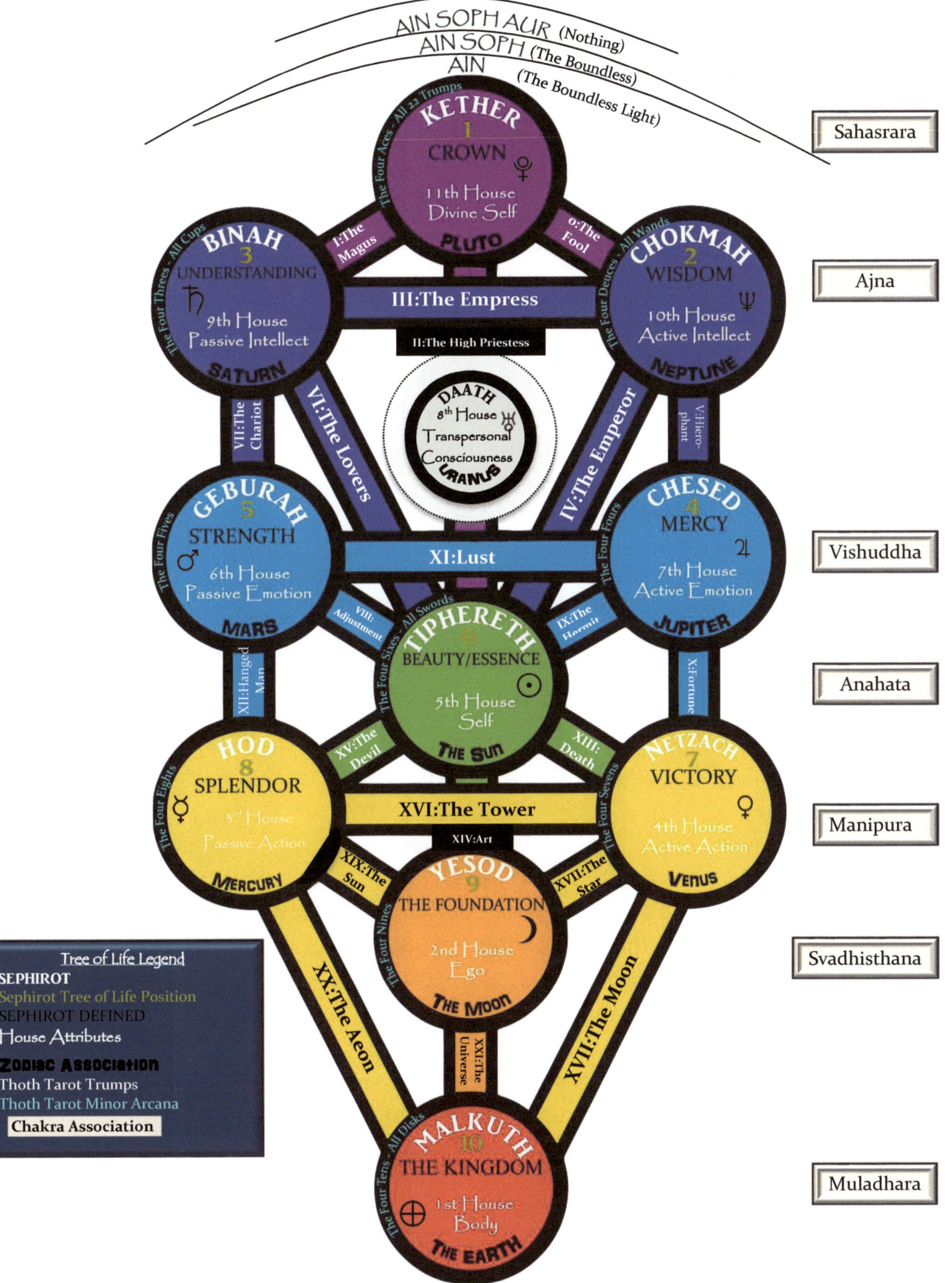

AIN SOPH AUR (Nothing)
AIN SOPH (The Boundless)
AIN (The Boundless Light)
The Four Aces - All 22 Trumps
KETHER
1
CROWN
11th House
Divine Self
PLUTO
I:The Magus
0:The Fool
The Four Threes - All Cups
BINAH
3
UNDERSTANDING
9th House
Passive Intellect
SATURN
The Four Deuces - All Wands
CHOKMAH
2
WISDOM
10th House
Active Intellect
NEPTUNE
III:The Empress
II:The High Priestess
DAATH
8th House
Transpersonal
Consciousness
URANUS
VII:The Chariot
VI:The Lovers
IV:The Emperor
V:Hierophant
The Four Fives
GEBURAH
5
STRENGTH
6th House
Passive Emotion
MARS
The Four Fours
CHESED
4
MERCY
7th House
Active Emotion
JUPITER
XI:Lust
VIII:Adjustment
IX:The Hermit
The Four Sixes - All Swords
TIPHERETH
6
BEAUTY/ESSENCE
5th House
Self
THE SUN
XII:Hanged Man
X:Fortune
XV:The Devil
XIII:Death
The Four Eights
HOD
8
SPLENDOR
3rd House
Passive Action
MERCURY
The Four Sevens
NETZACH
7
VICTORY
4th House
Active Action
VENUS
XVI:The Tower
XIV:Art
XIX:The Sun
XVII:The Star
The Four Nines
YESOD
9
THE FOUNDATION
2nd House
Ego
THE MOON
XX:The Aeon
XVII:The Moon
XXI:The Universe
The Four Tens - All Disks
MALKUTH
10
THE KINGDOM
1st House
Body
THE EARTH
Sahasrara
Ajna
Vishuddha
Anahata
Manipura
Svadhisthana
Muladhara
Tree of Life Legend
SEPHIROT
Sephirot Tree of Life Position
SEPHIROT DEFINED
House Attributes
Zodiac Association
Thoth Tarot Trumps
Thoth Tarot Minor Arcana
Chakra Association

The Sephirot, within the Tree of Life, corresponds to and have been shown to epitomize the ideals of many different faith based practices and rituals. In addition to Astrology and the planets, you will find that the ten Sephirot can represent and explain the Tao, the Gunas, the Chakras, Judaism, Thelema and many other belief systems; although, you may notice subtle differences in the spelling of some words and placement of certain planets. It can also be used to explain the human psyche and physical development, in terms of Enlightenment to Ascension.

Here is a brief example of a particular planet's alignment with a particular Sephirot. Accounting for evolution and illumination of consciousness, Pluto, the transformer, abolishes the old and creates the path for the new; fitting for Sephirot Kether, which is interpreted as absolute compassion.

Shiva Stones

Understanding how the Tree of Life's Sephirot reflect certain chakra portals allows us to determine which areas of our lives need cleansing and balancing; and the most effective means to accomplish that task.

If we find that Ajna (Third-eye) is weak, then our actions will reflect that we are lacking both wisdom and understanding. We have a weak connection with the Divine and rarely consider the consequences of our actions; short-sighted. Additionally, as the Sephirot Binah and Chokmah represent Understanding and Wisdom, they also, respectively, reflect your 9th House and 10th House. In terms of

Astrology, your 9th House is the house of travel, philosophy, and perspective. Your 10th House is the house that helps to characterize who we are to the World. It houses your ambition. This can also be interpreted to say that the 10th House contains your destiny or who you want to become.

An ill-defined Ajna shows that ones' actions reflect ignorance. They will experience the World viewing them as individuals without a future nor any plans of improving their current state of mind. They interpret those judgements as permanent characterizations of themselves and thus wallow in despair. If one could tap into the knowledge of this system, one would be able to immediately recognize that education is the key to resolving the imbalance of Ajna. The subjects they would need to focus on would be determined by their specific situation; which will always reveal the areas in which they are lacking expertise. Do not fall prey to hopelessness. Ask and ye shall receive, or you could always just Google it.

Agate Druzy Geode

If you absorb this understanding of the Tree of Life and incorporate it into your healing journey, you will not only open your eyes to the possibilities of physical health restoration but you will also learn of the many paths that exist for one to make desired positive changes a reality.

Sahasrara (Crown Chakra)

Sahasrara, pronounced Sŏh – hŏhs - rŏh – rŏh, in English is translated as "thousand-petal lotus". On the Tree of Life, Sephirot Kether is aligned with Sahasrara. Kether represents pure consciousness and union with Supreme Being, God, All Creator, Highest Vibrational Energy, or your identified representation of the highest "Form" in existence. In traditional Hatha Yoga, Sahasrara's Biju Mantra can be silence or the chant "Ng"; which is made audible by sounding out the word using the back of your throat. It is similar to the sound of "ng" in the word "king". Your Crown chakra is located near the very top of the head and hovers an inch or so above it. Sahasrara is illustrated by the color purple and white.[10]

When Sahasrara is out of balance or blocked, one can seem spacey, ungrounded and impractical. Some may be over-stimulated, causing indecisiveness and have difficulty completing tasks. You may often feel confused and develop delusional or grandiose ideas.

The over stimulation can physically manifest itself in problems with the brain such as, mental illnesses, schizophrenia, insomnia, anxiety and even headaches. People can have obsessive thoughts so frequently that basic life rituals and hygiene practices are ignored.

Possessing an under-stimulated Sahasrara can result and selfish behavior and even depression. Individuals feel cut off from spirituality; plagued by a sense of meaninglessness. You may feel a lack of joy or a lack of values and exhibit unethical behavior and thoughts. The weaker the flow of energy to your Crown, the more nonchalant you may become. This lack of life force to your intellectual center breeds feelings of unworthiness. Physically, because Sahasrara is associated with your brain stem and spinal cord, you may have trouble balancing or develop vertigo.

Sometimes the cause of the disruption in the flow of chakra is initiated from physical trauma. For example, a car accident resulting in a physical injury to the spine can affect your Sahasrara chakra balance. Balancing Sahasrara creates an emotionally centered, mentally healthy human being.

Agate Geode with Amethyst Center

Sahasrara is the one who balances and harmonizes the interior and exterior areas of our physical nature. It manages our spiritual center where we recognize the interconnectedness of all living things. Many may find it challenging to release the sense of self-preservation, me, myself, and I, as the center of their universe. If you can allow yourself to accept the life source from the Universe and the Divine, you will set the stage to realize your full potential in this life and your future life-times. Being balanced and aligned psychologically draws you to take an active leadership role in your existence, owning the impact you have on yourself and, regardless of its size, the part of this World you have created.[11]

Sugilite

Ajna (Third-eye)

Ajna, pronounced Ŏhj – nŏh, is the Brow or Third-eye chakra. Ajna aligns with two Sephirot, Chokmah (wisdom) and Binah (understanding). Recognized as the 6th primary chakra, Ajna in English means "beyond wisdom". It is illustrated by a transparent dark purple, indigo or dark blue lotus flower with two white petals. Its Biju Mantra is "Aum" or "Om". The chakra point is located in the brain, directly behind the center of the eye-brows.[12]

Ajna, if over-stimulated or too open to the energies of the Universe, can be depicted as having serious issues interpreting or perceiving your interconnectedness with the World. One will misunderstand another's intention and assume that the entire world misunderstands them. You will also have an overactive imagination on such a level that you create realities that are completely out of line with the current physical reality. The problem with the false realities is that they are born out of fear and confusion. Dangerously, one can become proud, dogmatic, and authoritative.[13]

If under-stimulated, Ajna blockage is represented by a lack of vision, lack of concentration, and quite frankly, a lack of imagination. You are unable to see the big picture with a clouded intuition. One may become extremely timid and fearful; overly reliant on logic and intellect. Manifested dis-ease is seen as hormonal imbalances due to its close proximity to the pituitary gland. One may also experience constant headaches, vision abnormalities, unexplained fatigue, frequent mood changes, and irritability.

Charoite

Ajna is our path to inner wisdom and the wisdom of life that transcends. Opening, developing, and balancing Ajna requires dedication and patience. Ajna is associated with intuition and knowing of the seemingly hidden. Mainstream society teaches us to fear such things or dismiss as dangerous, seductive folly. However, if one is to understand their existence, a true sense of Self must be realized. When Ajna is balanced, your senses are free to openly interact with the spiritual, emotional, and physical World; there is no resistance. In the same stance, your psyche is strong and protected from invading energies; whether they be negative or positive.

All intuition flows through Ajna. A Balanced Third-eye allows you to not only visualize the changes you would like to see in yourself and the World, but it also helps you to see the path necessary to harmoniously bring those changes to fruition. You become aware of your environment on a spiritual level. Both hemispheres of your brain function in unison; the right is creativity and the left is logic. You release stress and anxiety because, due to your insight, you are now 'Awake'. You gain greater control over your emotions. You conquer fear and can develop extrasensory gifts. Telecommunication is one such gift that some individuals cultivate. Other possible gifts are clairvoyance (seeing glimpses into other realms), clairsentience (knowing through feeling or sensing) and other types of psychic abilities.[14] All in all, you develop the ability to "see clearly".

Charoite

Vishuddha (Throat Chakra)

Vishuddha, pronounced Vĭ – shū – dŏh, is the 5th primary chakra. Its location is near the spine at the same level of the throat. Vishuddha is illustrated by a sixteen-petal lotus whose color ranges on a scale from light blue to blue-green. Vishuddha's Biju Mantra is "Ham". On the Tree of Life, its aligned Sephirot is Chesed (generosity) and Geburah (strength & power).

When Vishuddha is over-stimulated, one can exhibit arrogance, sarcasm, dogmatism and an overbearing attitude. Individuals find it difficult to verbally control expressing their thoughts. They become manipulative, inconsistent with communication, and deceptive of themselves or others.[15]

When your throat chakra is under-stimulated, extremely low energy flows; you may become physically weak and unreliable. You become unable to communicate ideas, have problems with self-expression (expression of your own truth). You do not know how to ask for what you need and you find it difficult to build the life you want. You may notice that you have developed a small, imperceptible voice, extreme secretiveness or shyness. You exhibit a loss of connection with your life's purpose. Under-stimulation also causes one to fear being judged or fear of not being accepted which, in-turn, causes one to withhold expressing their honest opinion.[16] [17]

Vishuddha's energy connects directly to your bronchial tubes, vocal cords, respiratory system and all areas of the mouth; including the tongue and esophagus. Physical dis-ease presents itself in the form of bronchitis, asthma, sore throats, swollen or sore tonsils, thyroid problems, shoulder and neck pain, hearing loss, jaw pain or TMJ, mouth sores, etc.

When Vishuddha is aligned and balanced, you easily attune to both inner and outer vibrations. You obtain contentment, peace of mind, a good sense of timing and develop a strong aura of faith. You believe in and stand by your convictions and harmoniously face opposition with consideration. Your sense of understanding and sympathy for others is increased. You will discern and recognize what is in your heart. You easily verbalize your desires and will understand how to be authentic with others and yourself. Once aligned, it is common for many to tackle repetitive trials in order to learn how to authentically interact with others. This also includes mundane tasks such as writing in a journal and practicing in front of a mirror. Speaking your mind and expressing yourself honestly does not mean you should be cruel and disrespectful. Truth from your spiritual essence will be presented with compassion and kindness.[18]

Blue Kyanite

Anahata (Heart Chakra)

Anahata, pronounced Ŏh – nŏ – hŏ – tŏ, is the 4th primary chakra. Translated in English from Sanskrit, it means "sound produced without touching two parts," "pure," or "clean". On the Tree of Life, its Sephirot is Tiphareth (beauty, essence). Visually represented by a green lotus with twelve-petals, its Biju Mantra is "Yam".

When Anahata is over-stimulated, we become possessive, demanding, moody and controlling. You remain in unhealthy relationships. Loving someone 'too' much is a sign of an overactive Anahata. You may find yourself constantly looking to rescue someone, be a hero, or committing to relationships simply to feel needed and desired.[19]

With a blocked or weakened Anahata flow, you are oversensitive, overly sympathetic and feel the need to constantly give yourself to others. If you find yourself constantly needing to have others' input, permission, and co-signage, then you are codependent; another sign of an imbalance in Anahata. Blockages can manifest as loneliness, lack of emotional fulfillment, and difficulty accepting love and kindness. There is an apparent lack of sense of connection to the Divine and to nature. You have unresolved sorrows, hold tightly to grudges and refuse to forgive.[20]

Imbalances in Anahata directly affects the chest, heart, and lungs. Issues can manifest in the form of asthma. The problems in your heart muscle can manifest issues with blood circulation. Along these lines, you may see symptoms of heart disease, irregular heartbeats, angina, high or low blood pressure, and weakened movements in your arms and hands.

Green Aventurine

When you experience true love through Anahata, it is from a Divine plane, high above the animalistic level which deals with romantic and sexual intimacy. This type of love transcends the physical realm and solidifies our state of Being across all realms, including the spiritual. To truly understand the essence, you must step outside of yourself and accept that love, in terms of Anahata, is the Essence, the Energy, the Life-Itself, that has the ability to transfigure our experiences and emotions. When balanced, we can face any situation with compassion, full of self-acceptance and respect for ourselves and others.

Being the source of joy and unconditional love, it reveals the truths that are not able to be expressed in words. Whether or not others receive you in the same way will not falter your resolve. Aligned, you will walk and communicate in an aura of spiritual understanding, connecting you harmoniously to everything and everyone you encounter. [21]

Rose Quartz

Nephrite Jade

Manipura (Solar Plexus Chakra)

Manipura, pronounced Mŏh – nee – poo – (r)ŏ [*roll the "r"*], is the 3rd primary chakra whose energy center is located half way between the base of the sternum and the navel. Its Tree of Life Sephirot are Hod (intellect) and Netzach (emotions and spontaneity). Hod is the actualization that controls, dismantles, and reconfigures energy into different forms. These forms may be in opposition of one another or perfectly balanced. This is where Netzach steps in possessing the quality of energy needed to overcome strong forces. The nature of Manipura is removing interference, creating room for your will to be brought to fruition. Manipura translates to mean "lustrous gem". In images, it is represented by a yellow lotus of 10 petals with shades of red, dark-blue, and black. While meditating, you may repeat is Biju Mantra "Ram".

When Manipura is over-stimulated, your behavior depicts a misuse of power, dominance, and over reliance on your will; which can potentially lead to physical exhaustion. These are traits of workaholics, perfectionist, and overly demanding individuals.[22]

When under-stimulated, one may see an increase in feelings of helplessness or that you bear no responsibility for yourself, your situation nor the people around you. You lack clear direction, purpose, or even ambition. You may even find that you will start many different projects only to never complete them. You exist believing you have no control over your own life; which sparks confusion and fear.

Citrine

As the powerhouse of the gut, Manipura's channel affects the intestines, pancreas, liver, bladder, stomach, upper spine and quite possibly surprising, the eyes. When balanced, you feel brighter and with the light of the morning sun, you see better. Physically manifested die-ease can be seen as digestive problems; such as irritable bowel syndrome; poor performing adrenal glands, diabetes, allergies, fatigue, dehydration and hypertension.

When balanced, our solar plexus chakra provides warmth, confidence, self- esteem, and true happiness from deep within our Being; not the shallow emotions associated with the fleeting satisfaction from material things and physical pleasures. This chakra allows us to sense our personal power and helps us to build our self-reliability. We are able to transform our will into action and movement; as it is also the center for our willpower, self-discipline and our perception of the Self.

Aligned, you are able to make choices consciously and act upon those decisions with confidence, creativity, and power. You can determine your life's purpose and move forward without stress and anxiety afflicting you. You are assertive, respectful, and effortlessly achieve your expected results.

- Citrine

Svadhishthana (Sacral Chakra)

Svadhishthana, pronounced Svŏhd – heesh – thŏ – nŏ, is the 2nd primary chakra. Your sacral chakra point is located at the level of the sacrum, in the small of the back; about two-inches below the navel.

Svadhishthana translates to indicate "where your being is established" or "home of the vital force". Its Biju Mantra is "Vam". Illustrated by an orange six-petal lotus, Svadhishthana is associated with the Sephirot Yesod (correlating with the sexual organs). Yesod's function in the Tree of Life is to collect the various energies that have been created in the descent of the tree, downwards, and distribute them to Malkhut. Malkhut is the material World, where energy can be physically manifested.[23] [24] For example, conception is the joining of opposing energies creating new life. This energy is delivered by way of Yesod to Malkuth to realize its birth or 'physical manifestation'.

If over-stimulated, we exhibit signs of distrust and aggressive traits of being overly self-indulgent; focused on physical pleasures. We become hyper-emotional, overly sexual and wracked with feelings of guilt. Out of balance energies will manifest addictions, both physical and emotional. You also tend to get lost in your fantasies about life's pleasures instead of actually living life.

If under-stimulated, one can be resentful towards the life they find themselves living and their chosen actions. One becomes full of anger, depression, and frustration; compounded by the fear of change. Emotional instability is a clear sign that there is an internal struggle for cohesiveness. You exhibit signs of a low-energy, low-libido, difficulty experiencing joy, and come off emotionally cold. You tend to hold back which bottles up the emotional load. As a result, you overload and your chakra stores now become over-stimulated. This is a vicious cycle that leads to irritability and problems with the reproductive and urinary systems. You could experience cramping, irregular menstruation and frequent infections; as the correlating organs are the bladder, prostate, ovaries, kidneys, gall bladder, bowel and spleen.

Carnelian (Orange)

When you are balanced in Svadhishthana, you will find it easy to relate to other people. You will be passionate, sensual, creative, connected to your feelings and fully present in your Being. You are able to let go and release your hold on what you assume to be "wholeness", and you allow yourself to freely feel change; which naturally occurs, constantly, within our bodies.

We have been "trained" to behave and told that emotional expression equates to sensitiveness, which is frowned upon.

However, a balanced Sacral Chakra helps us to reconnect and maintain that connection between our physical and emotional bodies, regardless of energy fluctuations.

Chalcedony (Orange)

Muladhara (Root or Base Chakra)

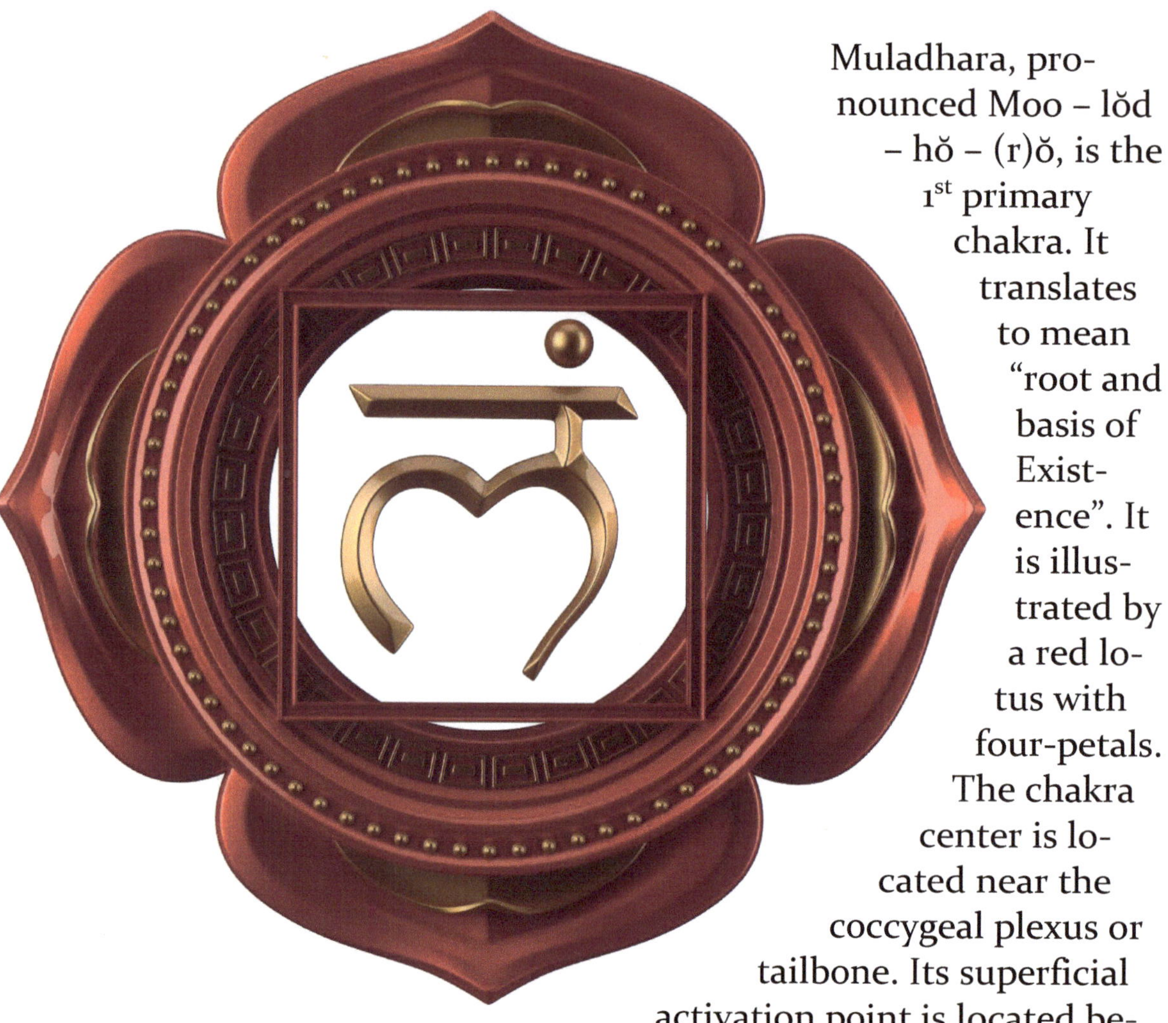

Muladhara, pronounced Moo – lŏd – hŏ – (r)ŏ, is the 1st primary chakra. It translates to mean "root and basis of Existence". It is illustrated by a red lotus with four-petals. The chakra center is located near the coccygeal plexus or tailbone. Its superficial activation point is located between perineum and pelvic bone. Muladhara's Biju Mantra symbol translates to "Lam". Lowest on the Tree of Life, the corresponding Sephirot is Malkhut. In the Kabbalah, Malkhut performs the same transcendental role as the basis of physical nature; creating, forming and moving creative energy. Physical manifestation of creative energy can be represented as conception and birth, dreams into reality, or grand events developed from a single idea.

To be aligned is to be grounded. When you are ungrounded in Muladhara, you are insecure and uncertain. Under-stimulation of your base plagues you with constant responses to the environment and human interactions using your fight or flight reflex. Nightmares, anxiety, constant uneasiness or other stress related disorders are the result of being ungrounded.

When Muladhara is over-stimulated, one can exhibit the signs of self-centeredness and addiction to money and sex. Whether surrounded by others or alone, one can feel smothered and antsy; desiring to get out or get away, escape.[25]

Physically, Muladhara affects your testes and other external sexual organs, your kidneys, and your spine. Over-stimulation leads to lethargy, colon problems or problems with elimination and cramps in the lower limbs. Some individuals manifest, eating disorders; as they can be a form or escape or comfort.

Red Jasper

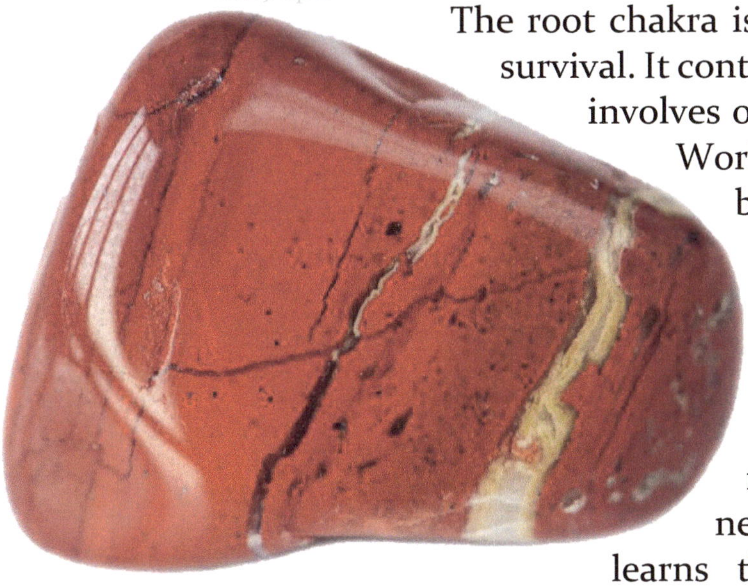

The root chakra is our instinctual center for survival. It controls our primal nature. This involves our primitive sensing of the World; which takes place as a baby enters this world called "rooting". This sensing is carried out as the infant finds food, feels secure, and sees and experiences a safe environment. If the primal needs are met, the Being learns that they belong to this World and will be supported. If your needs are not met, then as you develop, you inherently lack trust and a sense of security. As a result, you will find imbalances securely implanted in your root chakra.

Whether or not you feel secure in the present time is significantly impacted by how safe you felt as a small child.[1]

Root chakra, balanced, helps us to connect to the Earth as grounded Beings, providing the way for our energies to propel us forward through our life's journey. We develop improved concentration and clear thinking. We become courageous and feel vibrant, alive, full of energy, and driven. The root chakra increases our level of vitality and this helps us maintain our grounded state.

Muladhara is considered the densest of the seven chakras and most stimulating. The grounding of your energy sets the stage for balancing the other six primary chakras. Being aligned, you develop the ability to work through misfortunes with wisdom and transform them into valuable experiences and opportunities for development.[1]

Desert Rose
(Gypsum Selenite)

Chapter 3: Crystal Healing Therapy

To balance and clear blockages from our chakras, individuals can use any number of alternative methods. Meditation, yoga, prayer, positive visualizations, therapy (talk, sound, music) and Crystal Laying are practices that can be used to help you find your center and start living your Best Life. Focusing on Crystal Healing Therapy, the following list of stones can be used to balance one or more chakras. Non-invasive, yet truly effective, stones and crystals can be used at any moment of the day or carried with you at all times.

Aqua Aura Quartz

I will list nine crystals that can be used in unison to balance each of your chakras. In addition to the seven stones used directly to cleanse and balance your chakras, I will also explain the qualities and healing benefits of Clear Quartz Crystal and Black Tourmaline; who, for the purposes of balancing, will be positioned as spiritual and physical support anchors. During your healing session, Quartz Crystal will be placed at your head and Black

Amethyst

Tourmaline at your feet. This creates a complete grounded, protective circle. While cleansing and balancing, your energy will shift and may become even more unbalanced during an active session. Think of a cut filled with dirt. Initially, the pain is minimal or nonexistent. When you use water or other cleansers to flush it, suddenly, you feel sharp or stinging pain. No one ever said healing would be comfortable, but we do know it is necessary to elevate to a much higher, stable plane of existence. These two stones serve to keep you on track and in a state of progression during your healing sessions.

Malachite (Bullseye)

A Brief History of Crystals & Gemstones

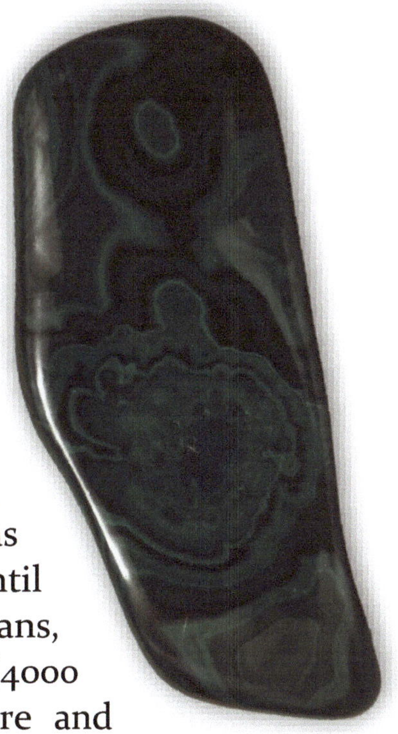

We understand that humans have been using stones in ritualistic practices since the Palaeolithic Mousterian period dated between 150,000 and 60,000 BCE. These stones were inscribed with symbols and letters denoting their purpose for magic, holy or sacred rituals.[26] During that period, stone beads were also used to fashion necklaces; and later, as amulets.

One of the initial gemstones identified, as being used for beaded jewelry and blades in tombs, was Carnelian. This identification did not take place until the Mesolithic (Middle Stone Age) Era. Sumerians, during the Fourth Millennium (4000 BCE), as noted in their lore and tablet inscriptions, paid great homage to the crystals and gemstones having great importance in society and their way of life. The inscriptions showed they believed gemstones had magical powers, making them useful for treating ailments, relationship issues and shielding ones,

Lapis Lazuli

possessing certain stones, from criminals.[27] Jasper was one such stone sourced by the Sumerians to comfort and ease pain experienced by women during pregnancy and labor.[28]

Aquamarine (Beryl)

Supported by records of account in the Antient Eqyptian Book of the Dead (Sir E. A. Wallis Budge translation), Eqyptians equated crystals and gemstones with the Eternal and Immortal. This inspired their dedicated practice of lining tombs and mummified bodies of Pharoahs with crystals, metals, and gemstones; such as Malachite, Turquoise, Carnelian, Gold, Lapis Lazuli and Jasper.[29] Of the same historical era, the Fourth through to the Second Millennium, the Old Testament reveals the importance of gemstones to those who followed the religion during this time period. The book of Exodus speaks of the breastplate to be made by Moses, in accordance with the Will of God, for his High Preistess brother, Aaron; allowing them to visually communicate with God.[30] The breastplate was to be crafted with twelve stones. Twelve is a sacred number signifying the Twelve Tribes of Isreal. Three gemstones would be aligned into four rows as follows: Ruby, Topaz, and Beryl; Turquoise, Sapphire and Emerald; Jacinth, Agate, and Amethyst; Chrysolite, Onyx, and Jasper.[31] In Ezekiel, of the Old Testament, a reference was made to the "Son of Man" being adorned with precious gems: Ruby, Topaz, Emerald, Chrysolite, Onyx, Jasper, Sapphire, Turquoise, and Beryl.

Encompassing instructions for the religious individual's way of life, the Old Testament teaches that the "wearing of the precious stones was only allowed in connection with the Divine", and wearing or using the stones for reasons of negativity was prohibited. In

Revelation of the New Testament, St. John illustrates that the New Jerusalem is surrounded by a wall built from twelve layers of precious gemstones. From the first to the twelveth layer sits Jasper, Sapphire, Chalcedony, Emerald, Sardonyx, Carnelian Chrysolite, Beryl, Topaz, Chrysophase, Jacinth, and Amethyst.[32]

Flowing through the Middle Ages on to the Modern Age Eras, we observed a shift from classifying crystals and gemstones only for their practical use to also adding its structure, chemical composition, atomical weight and other scientific class notations. This allowed humans to understand the use and radiance of these Universal treasures on many levels.

The healing properties of crystals and gemstones have been in practice and documented for years. They have been used to communicate with the Divine, facilitate healing, provide protection, represent royalty and placed on 'show' for their brilliance and beauty.

How Does Stone Healing Work?

This is the number one question I am asked. I feel that is one of the most important questions to be addressed, because it builds the foundation for understanding the physical and accepting the meta-physical aspects of stone medicine. What initially attracts the majority of us to a particular stone is its color. However, there is a deeper understanding to be had. Let us consider light, frequency, and our mind's eye.

Light is defined as an electromagnetic vibration at certain wave-lengths. Light produces color. For humans, wavelengths visible to the naked eye are between 750 and 380 nanometers[33]. All of these colors combined, within the spectrum, create what is identified as 'white light'. In a stone, its color appears as it does because it actually absorbs one or more colors of the spectrum. For example, a stone like Black Tourmaline has a black complexion because it absorbs all of the colors with in the spectrum. Understand that the mineral composition of the stone affects its color and sub-sequentially its weight. This color and mineral weight gives each type of stone a unique energy signature and vibrational frequency. This allows us to identify the type of stone. This is also what we "hear", literally and intuitively (via our mind's eye), when we are being called by or drawn to a stone. I use the word "hear" in quotations

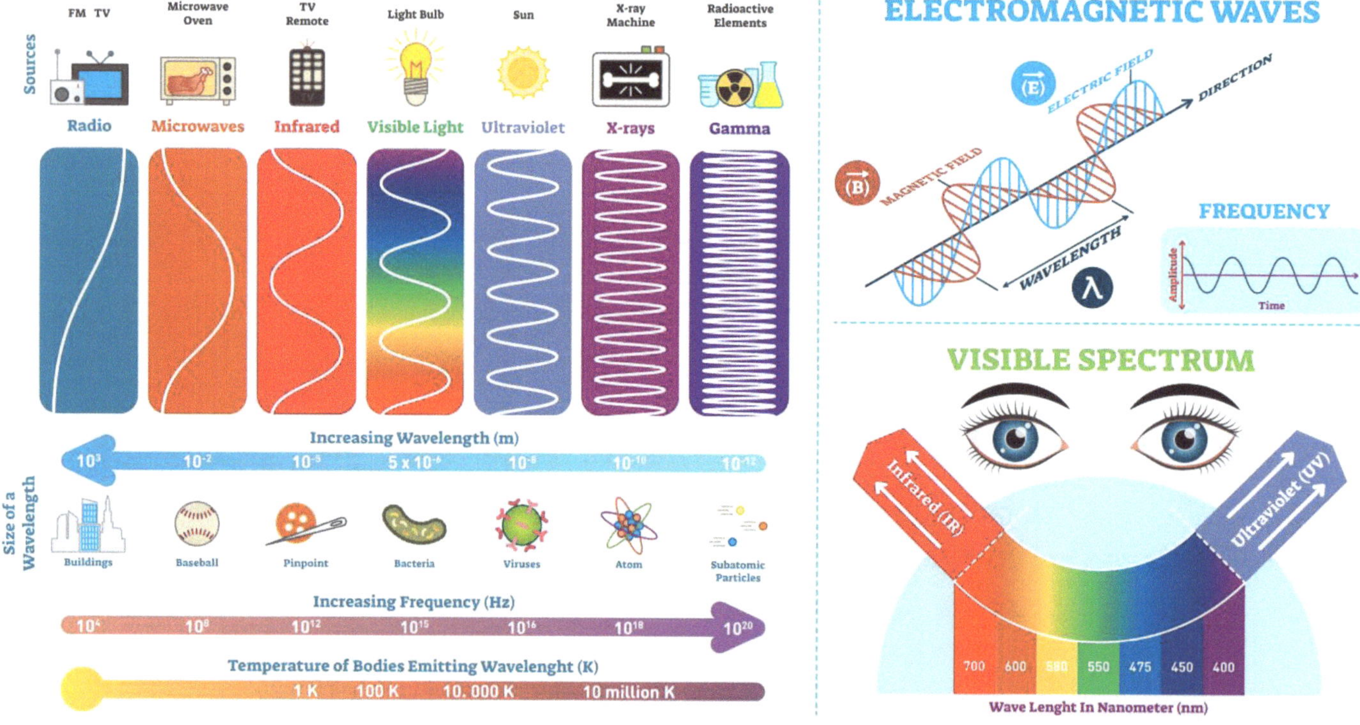

because it is the vibration of its frequency that calls to us. It is also this vibration that resonates with our energy signatures. On some levels, the "calling" invokes emotional responses. Through practical application and many years of study of mineral composition, we have come to understand the physical, emotional and spiritual healing aspects of crystals and gemstones. Metaphysical healing is not always reflected physically, but is usually felt in a spiritual sense or on an emotional level. Changes can be noted in attitudes, perspectives, developed intuitions and overall moods.

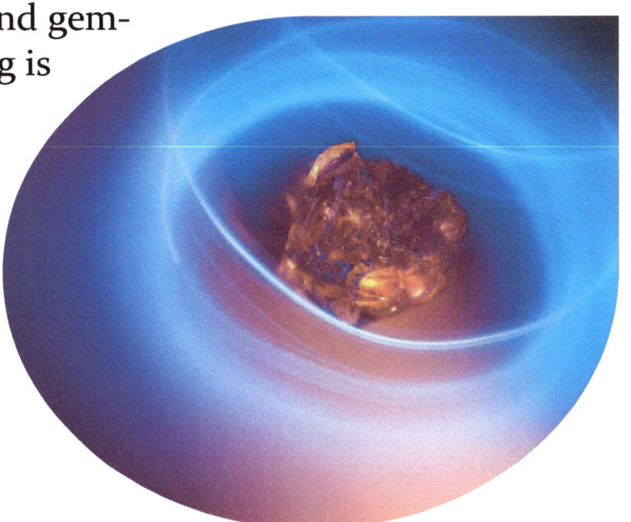

Scientists at the National Institute of Standards and Technology's (NIST) Center for Nanoscale Science and Technology, have also developed a technique for measuring crystal vibrations by using graphene. Graphene is a one-atom-thick layer of carbon atoms arranged in a hexagonal lattice.

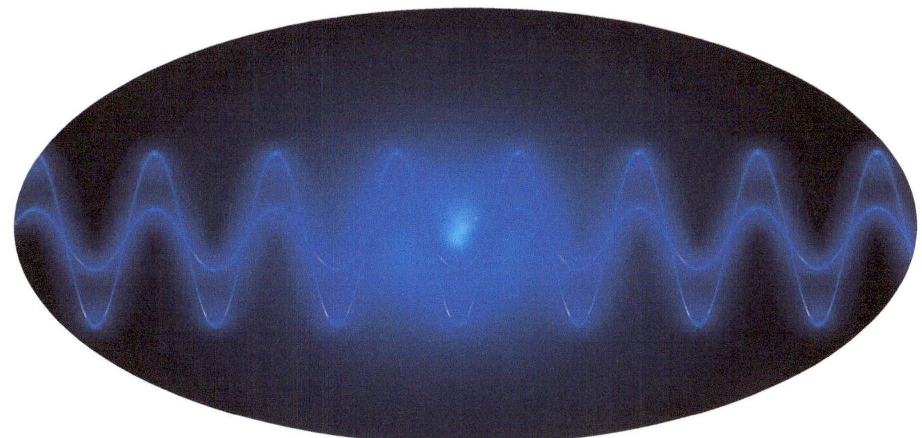

It is 200 times stronger than steel, an excellent conductor of heat and electricity, it absorbs light and is the thinnest material known to man. It has many uses, such as: batteries, computer chips, touch screens, etc. The researchers have used graphene to help measure the unique vibration of crystals and stones[34].

While it is true that humans cannot see certain wavelengths nor hear certain frequencies, we know for certain that our bodies are still greatly affected by them. Infrared, microwave, radio, ultraviolet, x-ray and gamma rays are invisible to the naked eye. However, if specific parts of the body have prolonged exposure to these wavelengths, we will sustain significant and even life threatening damage. This is one reason why cancer treatment, using x-ray radiation, is carried out in sessions over a predetermined period of time.

Understanding the composition of different crystals and gemstones helps us to not only accurately identify them, but it also helps us to determine why they are so effective in addressing certain parts of or even our entire Beings; Spirit and all.

Why Are Crystals Revealing Themselves, At Such A Rate, After Being Concealed For So Long?

Nature is seasonal. Growth happens in time. Destruction (new path paved) occurs to spur growth. If we apply this understanding to our healing, we can gain a different perspective. Our ancestry has led to the current structured civilization. We, the entire "World Community", must face a steep curve to obtain the necessary growth that allows us to begin living and walking the path of our calling, in the here and now. This steep curve can also be interpreted as "Kung Fu" or the putting forth of great energy, dedication, and time to learn and develop a skill. The newly paved path, or destruction, is made to our old ways of thinking. Our current path, ignoring our impact on our fellow man and the environment, is not sustainable. It pushes us out of season.

Labradorite

For example, we eat certain foods year round. This is not how our ancestors lived and maintained harmony with the environment. Eating foods out of season means we are manipulating nature to grow out of season. We are throwing nature off balance and something has to be taken from another to correct that external imbalance. Thus our bodies will then need to take in the energy needed to correct the internal imbalance. Nature has a funny way of recovering and thriving in spite of our constant barrage of abuse. However, if we want to ensure our healthy survival alongside of Mother Nature, a great shift in thinking and change in actions of humans must take place. Many individuals state they are unsure of what to do or where to start. In order to heal the external, we must start internally. The Self must be made Whole.

Theory of Thumb: If you are living in the past, you tend to be depressed. If you are living in the future, you tend to be anxious and will likely experience frequent panic attacks.

Healing crystals were unearthed because there is a great need for them. Crystals and stones were developed to heal, inspire, and enlighten. This is their purpose. It cannot be fulfilled without human intervention. It aligns perfectly with the times. As we have witnessed, human beings have never been in a more dire need of assistance. As we have been destructive to the health of our Beings, it is reflected outwardly in how we treat one another and the environment.

To make lasting, effective changes, we need to address the inner turmoil. As we walk, talk, and interact, we emit our unique energy signature along with the vibrations that reflect our current state of Being. This frequency will attract and, in some cases, repel the frequencies of healings stones. It is this attraction to and repulsion of certain stones that provides insight into which aspects of our Being needs to be addressed.

Our energy signature is as unique as our fingerprints. When we conduct our daily or even while sleeping, our minds are actively emitting thoughts or brain waves. These thoughts are as a result of our past, present, and our possible future experiences and past-life experiences.

Lepidolite Slab

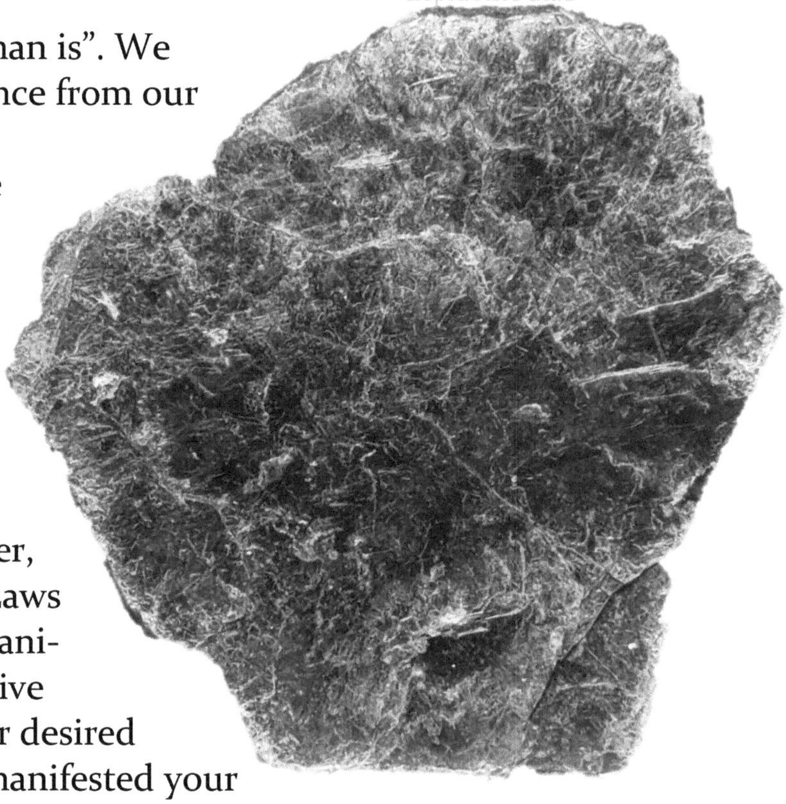

"So, a man thinks, a man is". We literally create our existence from our thoughts. The trouble is, many of our thoughts are formed with negative tones. These thoughts, or vibrations, physically manifest in our environment and within our bodies.

Consider this, using prayer, positive thinking or the Laws of Attraction, you may manifest a nice car. Your positive thoughts concerning your desired mode of transportation manifested your reality. However, your nice car is being paid for by a minimum wage job; which is great for social status reasons, but places strains on the finances. This stress, emotionally and mentally, could manifest as hormonal imbalances or heart muscle and breathing problems. This manifestation is the reality encompassing your current state of Being. Therefore, your unique energy signature, coupled with the vibrations of your current state of Being, forms the frequency you emit. Are you giving off good vibrations?

Stone medicine comes into play the moment you interact with a stone and have the desire to touch, hold, or just be in its presence. This is as a result of your energy field, emitting your frequency, creating an attraction based on your "need-for-balance". The gemstone(s) you are attracted to fits like a missing puzzle piece. Your wave pattern is dangerously radical or unremarkably undetectable and the gemstone's frequency centers and balances your energy. It works by emitting its frequencies, integrating with yours, and drawing other balancing energies to you or blocking interfering energies; so that you may find your own balancing frequency; thus strengthening your resolve and eliminating the unfavorable manifestation(s).

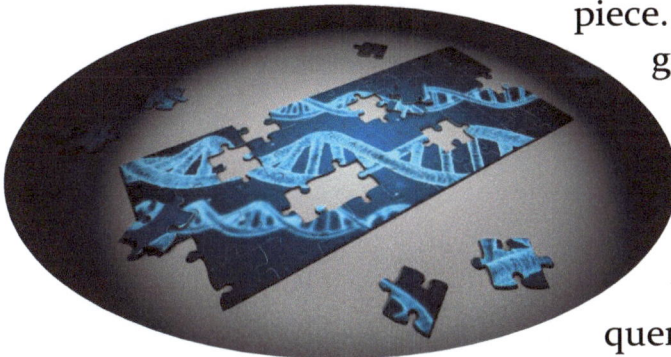

Consequently, if your car is repossessed or totaled in an accident, step back and assess the situation through a wide-angled lens. One may conclude that the car added so much negative energy that in bringing you to balance, it needed to be removed from your journey. The same applies to having a falling out with friends or relatives. Be open to changes in perspectives when conducting healing using crystals and gemstones.

Meditation

There are no special skills required for daily meditation. Additionally, there are not any required steps to take to get to your relaxed state. Meditation can take place at any time of the day, anywhere. Many of us fail to recognize that even while walking, we can reach a meditative state. Long-distance runners are one group of people who often reach meditative states while physically active. Performing repetitive tasks can take the pressure off of the "active" mind. You become autonomous and fall into a state of "flow". You can even reach this state while washing dishes or walking the dog. This state can be the break your mind needs to 'catch your second wind', allowing you to continue to push through your task. Meditation is that break our

minds need from the bombardment of everyday noises and interruptions. Using crystals while meditating guides and enhances the purpose of your experience, whether it is for healing a part of your body, building an aura of protection or cleansing your spirit.

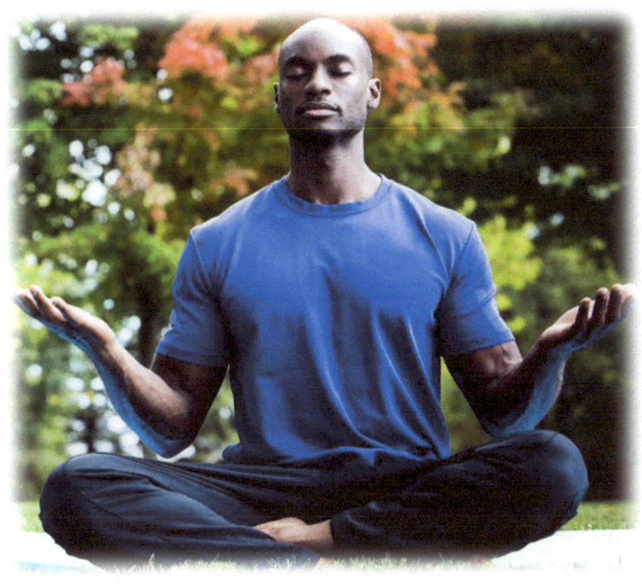

Prepping for healing calls for deep meditation. This practice allows your conscious mind to actively access the Super Conscious realm. This Space is also identified as the Spiritual Realm, Consciousness of God, or Angelic Realm. In this realm, you have the ability to reprogram your mind, open up to past lives, receive insight from the collective Super Conscious, draw the power to heal others and so much more. The length of time you are able to maintain your connection depends upon your ability to remain in a calm, quiet state of mind.

Essentially, it helps to try to reach a point in your mind where thoughts do not flow, just "empty space". It takes much practice to get to this level.

There are a number of techniques one can use to accomplish this. Experiment and find what works for you. One such technique is to breathe deep and often. You may visualize and concentrate on a particular color or object. Try not to visualize a living Being. Beings, regardless of the species, are tied to certain emotions. To reach a state of "empty space", it may be easier to avoid images that conjure emotions. Focus on your object or color, etc. until your thoughts slow to a crawl. In an empty space setting, your energy becomes neutral, providing the platform for the necessary changes to take place.

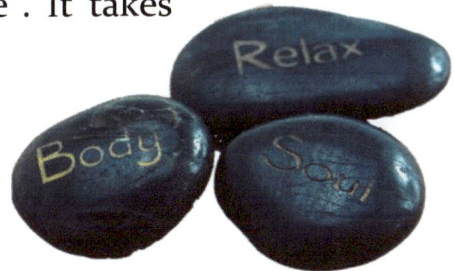

For healing sessions, beginning with neutral energy is ideal, as it allows for a more flexible session. During sessions your emotional and mental state may cause you to recall past memories. This is completely normal. Acknowledge those thoughts internally or address them aloud. This could be the key to a breakthrough. You can tailor your space, your time, your crystals used, and techniques. The important thing to remember that this is your healing journey and it will not be identical to anyone else's journey.

Your sessions should be conducted when you have time to yourself. Healing of this type is best done at regular intervals, e.g. once a day, every other day, or once a week. For beginners, it may be more beneficial to start by committing to 5-minutes a day. Increase your time by one-minute each day. If you decide to commit to one session a week, then double your time each week until you can reach a 20 to 30-minute session. You are more than welcome to increase your session times to one hour. The routines will help to build the habits necessary to maintain your practices.

Blue Sapphire

Meditation, as does healing, takes time. It is easy to talk yourself out of something you are not familiar with, even if the benefits far outweigh the use of energy dedicated. However, it should not be a chore. Make your path to personal healing your number one priority. As you heal, your positivity and healing energy will spread to those you care for and also to those who just happen to cross your path. Your entire World benefits!

Crystal Cleansing and Charging

Many crystals and stones can be cleansed by holding them under running water. This act helps to neutralize the energy they hold. As I explain the attributes of the following nine stones, I will also provide information on how to cleanse them. Unless otherwise stated, they will only require rinsing for 30-40 seconds to completely neutralize the unwanted energy within them. In addition, rinsing can drain the personal energy of your crystals and stones. This is why cleansing is normally followed by charging.

Some pieces require daylight, moonlight, Quartz Crystals, Selenite or Hematite to charge, just to name a few. One may also charge their stones by burying them within the Earth. Sunlight is amazing for charging crystals and stones, however, many are delicate and can fade when

Orange Quartz, Lemon Quartz, Strawberry Quartz & Blue Quartz charging on a Selenite Log

placed in the Sun's direct rays. Amethyst and Rose Quartz are two crystals whose color fades if placed in direct sunlight. It will not diminish their healing capabilities, however, different colors vibrate at a certain frequency. Changing the color changes the stone's frequency; which can change the type of healing effect it carries, not its healing strength. Along with the instructions given, you are encouraged to use your instincts to determine what will work best for your crystals. At minimum, most stones will need to charge for 2-hours. The optimum charging duration is 24-hours.

Cautions and Warnings

Green Opal

As they are comprised of certain minerals and substances, some gemstones are toxic. Crystal and gemstone toxicity is not usually discussed in areas of stone laying therapy due to the fact that it poses little risk if you are not ingesting the stone nor consuming a liquid that is made from direct contact with the stone. Even so, there are some clients who may feel light headed, antsy, or nauseous in the presence of certain stones. Energy sensitivity is not to be ignored. As you gain more clarity through your healing journey and begin adding stones to your collection, pay close attention to what you are feeling while in session.

Consider this, if you feel nauseous while addressing Svadhishthana, and you have a stone on your abdomen, this might be your body rejecting the idea of balancing your chakra. We can subconsciously push assistance away when it makes us uncomfortable. In some instances, this may indicate that you need to either shorten the time you are currently working with the stone or use a different stone for that chakra all together. In any case, you should immediately remove the stone in question. Wait at least an hour after the symptoms subside before trying again. Use the stone in short intervals, five to 15-minutes at a time. This will allow you to safely test your limitations while continuing to receive the benefits of working with your stones; at least until you find an alternate crystal or stone with which to work. If there are known cautions, I will list them for the 9 stones to be used to balance your chakras. For those new to stone medicine, it may be more beneficial to work with an experienced Crystal Healer to give you the opportunity to both enjoy and assess the experience before conducting self-healing sessions.

Associated Zodiac Signs

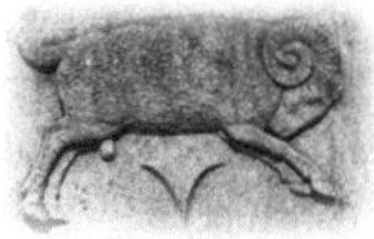

We acknowledge that as planets align, shift, and realign Earth is consistently barraged with their subsequent energies. This energy affects every living, animate and inanimate object within its reach. The energy, during certain cycles, is felt more strongly than others; which further supports zodiacal personality categorization.

If you are familiar with the zodiac signs, then you are aware that, although unique, we individuals, of the same sign share common traits, behaviors and can even react similarly to given situations. For these reasons, certain emotional and physical conditions can develop within our Being. Crystals and stones possess unique energies that have been identified to relieve and treat specific 'dis-eases'. They are nature's correctors and balancers. They contain the vibrational patterns that help elevate us to our highest

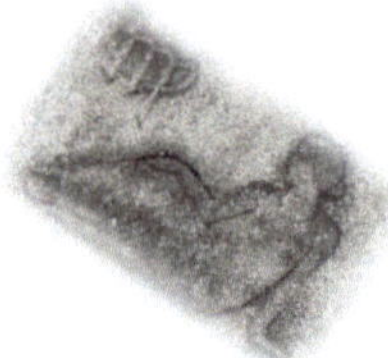

level of enlightenment. These stones have been noted to work impeccably with those born under a certain astrological sign; individuals which inherently present certain traits or particular conditions.

In this text, you will find, along with other attributes, the Zodiacs associated with each stone; amazing institutions that assists in gaining the much needed insight for self-development.

Although, I wrote this text, assigning certain stones to balance certain chakras, understand that Crystal Healing Therapy is comprised of stones that are able to be used to balance multiple chakras. Read each attribution and pay attention to your inner voice, your intuition, while looking at or holding each stone. You may feel led to apply them to chakras other than the ones suggested. Ultimately, this is your path to becoming a more wholesome and enlightened Being. Follow your instincts and remain open to the knowledge and healing to be gained.

Pink Tourmaline in Quartz

Green Fluorite

Chapter 4: The 9 Crystals for Complete Healing

Crystal Healing can be conducted with one gemstone or multiple stones. Crystals and gemstones can be used in conjunction with all other forms of alternative healing. Obtaining a diagnosis is one of the most effective way to begin your energy balancing journey. Understanding your symptoms and your desired outcome allows you to choose the gemstones that would work best to address your physical, emotional, mental, or environmental conditions. Many stones will address similar symptoms, however, they may be more efficiently attuned to a specific chakra; for example, as noted by color. With that in mind, take the time needed to conduct the necessary research to ensure you are incorporating the best stones for alignment. For your initial session, it might be more helpful to consult with your local Crystal Healer for insight and awareness.

As you continue to conduct regularly scheduled sessions, you will experience a shift in how your energies need to be aligned, this will inevitably, be as a result of symptom mitigation. At that time, you would, again, reevaluate the conditions and review the current gemstones in use. Map out a new configuration. Researching is efficient, but as you work with your treating stones, trust your instincts and inner guidance. You will not only see, but you will fully understand why and how you are experiencing the positive effects associated with Crystal Healing Therapy.

Quartz Crystal (For Completion)

Quartz Crystal, the salt of the Earth. Its six-sides symbolize the six chakras, with the termination or point of the crystal representing Sahasrara; "that who connects with the infinite". Clear Quartz Crystals reveal that the material plane (physical universe, things we can touch) can and does reach a state of physical perfection. They are a symbol of coming into alignment with cosmic harmony. Like pyramids, they channel high frequency energy into the physical ground of Earth. Quartz is able to simultaneously vibrate its energy at all of the color frequencies and all seven chakra centers; maintaining perfect alignment with the Light Divine.[35]

Zodiac Association:	All
Chemical Composition:	Silicone Dioxide (SiO_2)
Common Places Found:	United States, Europe, China, Russia, Madagascar, Himalayas, and South Africa
Chakra Association:	All

Properties of Healing

Chakra balancing is enhanced when you incorporate Quartz Crystal. Quartz's white light properties work with all other crystals and stones to enhance their healing abilities. Quartz provides clarity and can be used to remove negative thought patterns from the mind. These crystals are programmable. You can imprint your will, desire,

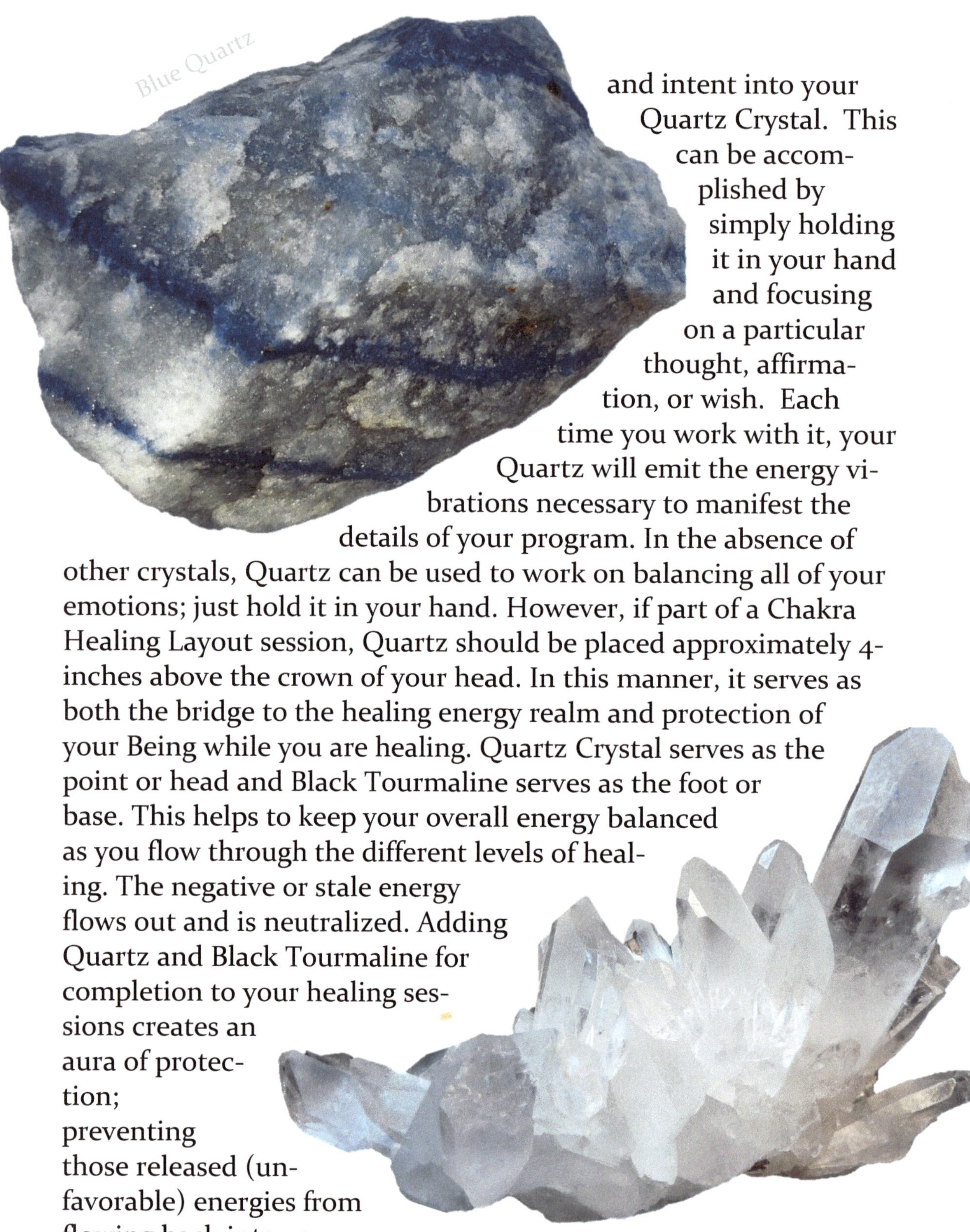

and intent into your Quartz Crystal. This can be accomplished by simply holding it in your hand and focusing on a particular thought, affirmation, or wish. Each time you work with it, your Quartz will emit the energy vibrations necessary to manifest the details of your program. In the absence of other crystals, Quartz can be used to work on balancing all of your emotions; just hold it in your hand. However, if part of a Chakra Healing Layout session, Quartz should be placed approximately 4-inches above the crown of your head. In this manner, it serves as both the bridge to the healing energy realm and protection of your Being while you are healing. Quartz Crystal serves as the point or head and Black Tourmaline serves as the foot or base. This helps to keep your overall energy balanced as you flow through the different levels of healing. The negative or stale energy flows out and is neutralized. Adding Quartz and Black Tourmaline for completion to your healing sessions creates an aura of protection; preventing those released (unfavorable) energies from flowing back into you.

When meditating with Clear Quartz, place it above Sahasrara. It should be about four-inches from the crown of your head. It does not need to touch your head, however if it does, that will not diminish the healing session.

Cleansing and Cautions

To cleanse Quartz Crystals, hold them in running water. They can be recharged by a number of methods. Direct sunlight, Hematite, Amethyst, Selenite, daylight and moonlight can be used to charge Quartz Crystals. Quartz Spheres and even clear Quartz can become a fire hazard if placed in direct sunlight. The sphere shaped crystal causes it to take light in from all angles; focus it to a point and deflect it out. Depending upon the angle, it can cause the point to hit a tree, the grass, or your home, resulting in a fire. It is very dangerous to charge those types of crystals in direct sunlight. To be on the safe side, just avoid direct sunlight charging for all forms of clear Quartz Crystals.

Tibetan Quartz

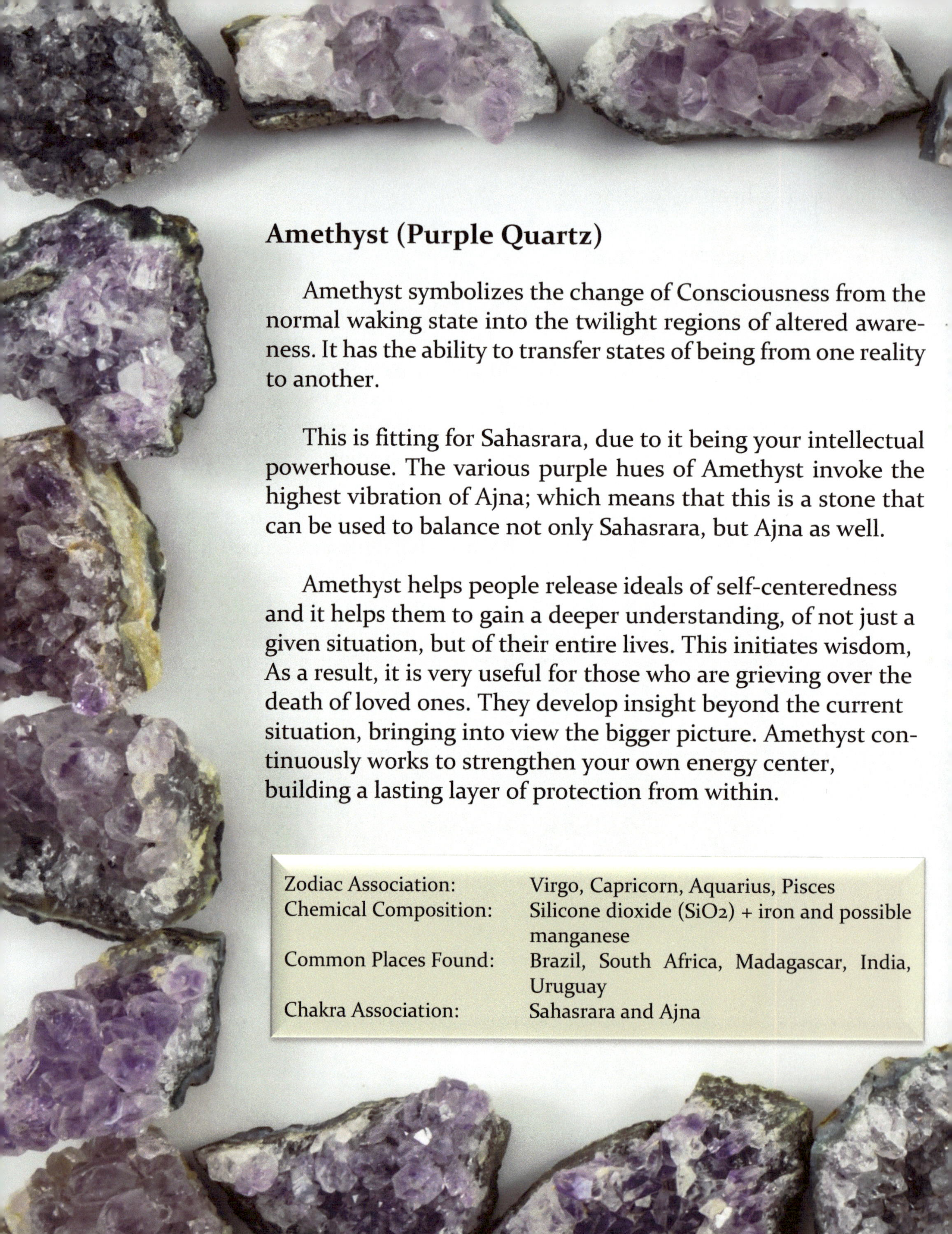

Amethyst (Purple Quartz)

Amethyst symbolizes the change of Consciousness from the normal waking state into the twilight regions of altered awareness. It has the ability to transfer states of being from one reality to another.

This is fitting for Sahasrara, due to it being your intellectual powerhouse. The various purple hues of Amethyst invoke the highest vibration of Ajna; which means that this is a stone that can be used to balance not only Sahasrara, but Ajna as well.

Amethyst helps people release ideals of self-centeredness and it helps them to gain a deeper understanding, of not just a given situation, but of their entire lives. This initiates wisdom, As a result, it is very useful for those who are grieving over the death of loved ones. They develop insight beyond the current situation, bringing into view the bigger picture. Amethyst continuously works to strengthen your own energy center, building a lasting layer of protection from within.

Zodiac Association:	Virgo, Capricorn, Aquarius, Pisces
Chemical Composition:	Silicone dioxide (SiO_2) + iron and possible manganese
Common Places Found:	Brazil, South Africa, Madagascar, India, Uruguay
Chakra Association:	Sahasrara and Ajna

Properties of Healing

Amethyst has a strong effect on the body, in terms of effectively addressing and working through high emotions and physical pain. As it is an effective tranquilizer, it helps to calm and quiet the mind.

Amethyst is also beneficial for fostering and maintaining a healthy immune system. It regulates hormones, helps to improve hearing and the symptoms associated with arthritis. Its energy vibrations have the ability to detox and cleanse the blood. This assists with mitigating bacterial infections, blood clots, viruses, acne, and other conditions that result from toxins building up within the body. Amethyst can also help the body fight the lure of addiction to drugs and substances. For those who suffer from addictions are fully aware that the urge exists not only in the body but the mind as well. Amethyst functions in the higher chakras, revealing its ability to merge the physical and mental healing planes. Thus it becomes a powerful tool for alleviating ailments that bring affliction simultaneously upon the body and the mind.

Regulating the body's internal fluids, Amethyst can bring relief to dry mouth, lips, and nasal passages. Cleaning the blood and increasing moisture, Amethyst helps the body to actively heal physical wounds.

Spiritual

Amethyst shows the mind how to surrender at the altar of "Thine Self". Only then can one cross the threshold into the realm of true knowing and wisdom. In this state, mental exhaustion and anxieties are released. This results in having the feelings associated with being overwhelmed, stressed, and overworked neutralized. This allows you to open Sahasrara, connecting your physical to your higher self; creating your best balanced existence.[1]

When meditating for healing with Amethyst, place the stone at the top or on the crown of your head. If lying down, place it to where it is touching your head at its center line.

Cleansing and Cautions

To cleanse Amethyst, hold it under cool running water. Amethyst has the ability to charge other crystals and stones. This is an indicator that Amethyst is capable of self-charging. Any rock crystal, such as Selenite, can also be used to charge Amethyst. Daylight is a welcome and strong source of energy. Just be sure to keep it out of direct sunlight. Amethyst has the tendency to fade in the Sun.

If experiencing dampness, a heaviness in the body, after you begin working with Amethyst, stop use immediately and wait until symptoms subside. Although these symptoms are rare, it serves best to be aware. Dampness can be characterized as an overabundance of water, i.e. swelling in the tissues. After the symptoms have subsided, reintroduce Amethyst in short intervals of 5 minutes at a time, each session, until you have maximized your contact time while maintaining the optimum level of comfort.

Additional Gemstones that can be used to Cleanse & Balance Sahasrara

Ametrine, Clear Quartz Crystal, Sugilite, Tourmilated Quartz and Purpurite.

Ametrine

Charoite

Charoite may not be one of the most well-known of the healing stones, but it is extremely powerful for psychic protection. Intuition or psychic protection is important for those who are going through the healing process and for those who conduct healing sessions. This applies to both clients and servicers, patients and doctors, young ones and mentors, etc. As you balance Ajna, your mind is susceptible to minute energy fluctuations. While that vulnerability exists, protection ensures that you will prevent invading energies from manifesting in your realm; allowing you to get the most out of your healing experience.

Zodiac Association:	Sagittarius and Scorpio
Common Places Found:	Russia
Chemical Composition:	hydrated potassium sodium calcium silicate hydroxy-fluoride ($K(Ca;Na)_2Si_4O_{10}(OH;F)\cdot H_2O$)

It can also be found to contain barium, aluminum, iron, manganese and strontium. Its calcium and fluoride content brings its effects to the bone level. The fluoride, as a halogen, in combination with the hydroxy groups purging of fact, helps drop poisons from the body. The electrolyte content supports fluid metabolism and conducts Qi or moves life force energy throughout and around the physical Being.[1]

Chakra Association:	Ajna

Charoite is named after the place from which it is mined; near the Chara River in Russia. Predominantly purple, it also has white, lilac, black, and brown inclusions.

Properties of Healing

Charoite is beneficial for reducing body temperature, as it draws heat. It also slows and regulates the heart rate. As another of our amazing detoxing agents, it works to repair and relieve the strain placed upon organs and systems within the body that have been overloaded with toxins. These include, but are not limited to, the liver, spleen, kidneys, digestive tract, the brain, and pancreas. Mental and social disorders are also treated and mitigated; such as ADHD (attention deficient hyperactivity disorder) and obsessive compulsive disorders.

Spiritual

Spiritually, its protection brings vibrations that empower you to merge the chakras of the Crown and Heart. It provides a Soul connection that assists you in moving forward through your journey of enlightenment. This stone can connect you with the energy of the Divine source. Divine source can be identified according to your walk in this lifetime, however it is the same source across all realms. Individuals identify it as a god, an angel, spiritual energy, etc. The source will strengthen the flow of your energies and all new spiritual gifts that come to light. It will anchor the Divine's light into your Being, using the chakra stores as points of entry and manifestation.

When meditating for balancing and healing, place the stone on your Ajna center.

Cleansing and Cautions

To cleanse Charoite, hold it under running water to remove all other associated energies that may have manifested in or around the stone. Afterwards, it needs charging to replace its energy reserves.

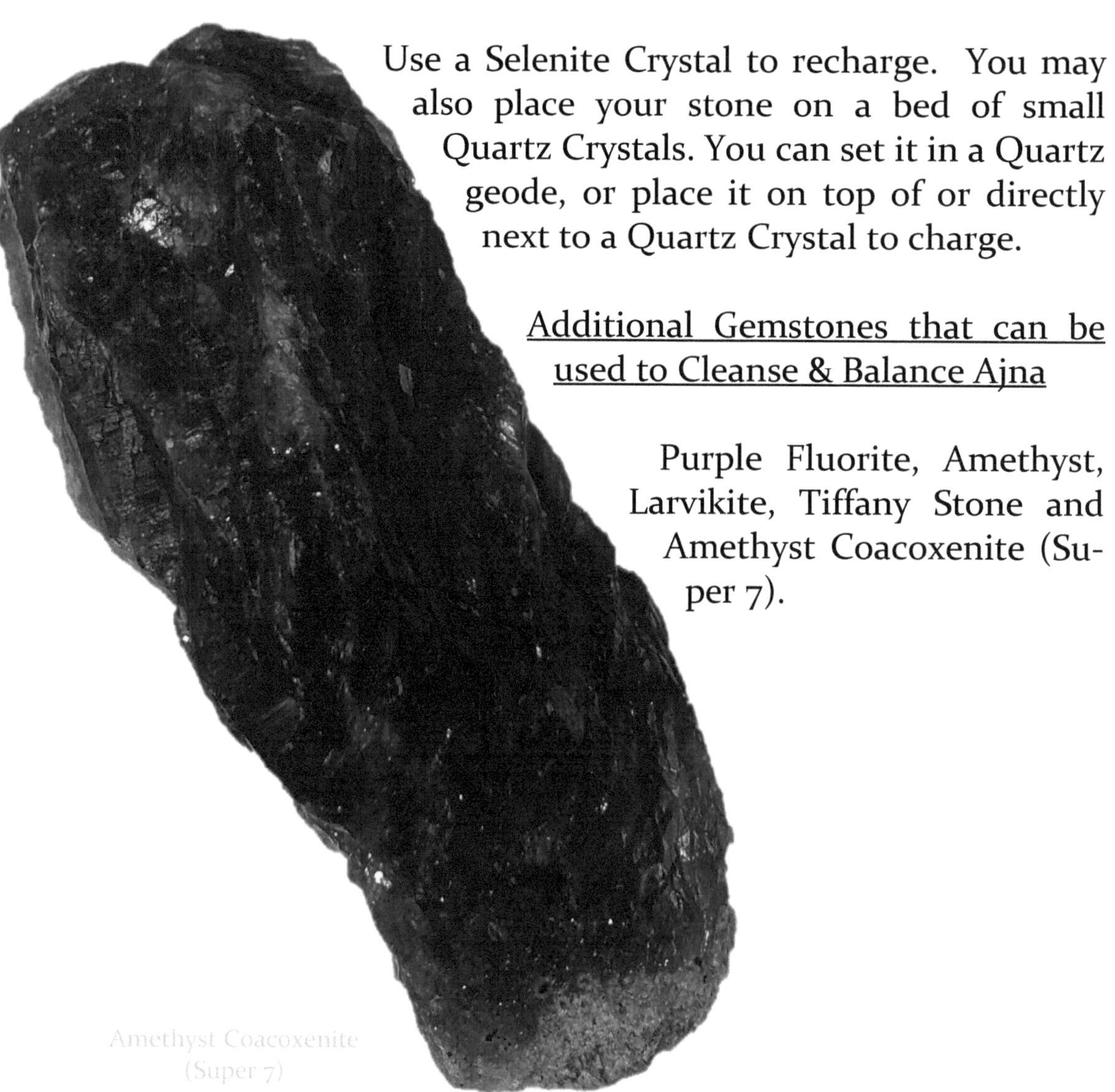

Amethyst Coacoxenite (Super 7)

Use a Selenite Crystal to recharge. You may also place your stone on a bed of small Quartz Crystals. You can set it in a Quartz geode, or place it on top of or directly next to a Quartz Crystal to charge.

Additional Gemstones that can be used to Cleanse & Balance Ajna

Purple Fluorite, Amethyst, Larvikite, Tiffany Stone and Amethyst Coacoxenite (Super 7).

Blue Kyanite

Blue Kyanite is a mineral that creates pathways due to its high vibrations and energy transfers. This stone helps to open mind Centers; enhancing telepathic and psychic abilities which are all means of communication. Kyanite is calming to the body and helps us to reach deep meditation. Due to its ability to rapidly exchange and transfer energy, Blue Kyanite will not absorb nor hold on to negative energy. Therefore, it can be used to cleanse and balance other stones just as effectively as it cleanses and balances Vishuddha, your Throat chakra.[36] [37]

Zodiac Association:	Aries, Libra, & Taurus
Common Places Found:	Brazil, Switzerland, Russia, Siberia, India, Kenya, & the United States (North Carolina & Georgia)
Chemical Composition:	Aluminum silicate (Al_2SiO_5)
	The aluminum is cooling and soothing to the nervous system. Kyanite can include trace amount of chromium, titanium, lithium, magnesium, manganese, and vanadium. The blue color is caused by a trace amounts of iron and possibly chromium. The green color includes both iron and the vanadium. The orange color includes manganese.[1]
Chakra Association:	Vishuddha

Properties of Healing

Blue Kyanite is a principle stone for relaxing the liver and autonomic nervous system. It positively affects the brain and spinal function. Concerning neurological conditions such as epilepsy, Blue

Kyanite helps to regulate and ease signals on the nerve pathways that lead to the brain and brain stem. It also eases seizures, convulsions, nervousness, spasms, twitching, symptoms of Parkinson's disease, temporomandibular joint disorders (TMJ), teeth grinding, poor motor sensory coordination and Tourette's syndrome. It also helps to relax the throat muscles and eases tensions with all forms of communication.[38] Its relaxing effects makes Kyanite an effective stone for dealing with hypertension.

Spiritual

Blue Kyanite

Blue Kyanite strengthens both your Ajna and Vishuddha functions. It helps one to manage, prioritize, and organize their feelings, thoughts, and emotions. Understanding your own feelings not only paves the way for effectively communicating your thoughts and emotions to those you care for, it also allows you to create meaningful , lasting relationships with others.

When meditating with Blue Kyanite, place it at Vishuddha, in the groove, about an inch below the Adam's apple.

Cleansing and Cautions

Cleanse Kyanite in warm water and recharge it using Selenite Crystals. You may place it in a Quartz geode or cluster. Kyanite has metal inclusions. Take care when formulating elixirs or gem water using this stone.[39]

Additional Gemstones that can be used to Cleanse & Balance Vishuddha

Blue Apatite, Turquoise, Blue Lace Agate, Lapis Lazuli, and Aquamarine.

Green Aventurine

Green Aventurine is applicable for any kind of physical and emotional healing. Universally in tune with Anahata, it is fitting that this stone helps to renew, restore, build, and make secure. Not only is it a healer, it is also a protector. It is an essential stone for neutralizing environmental pollution and minimizing their effects on our systems. Green Aventurine is an excellent stone for promoting and attracting abundance into your life. As it is strongly connected to the chest and thymus, it is an effective regulator and it works to normalize and harmonize your body's physical and spiritual systems.[40]

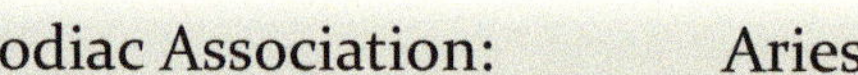

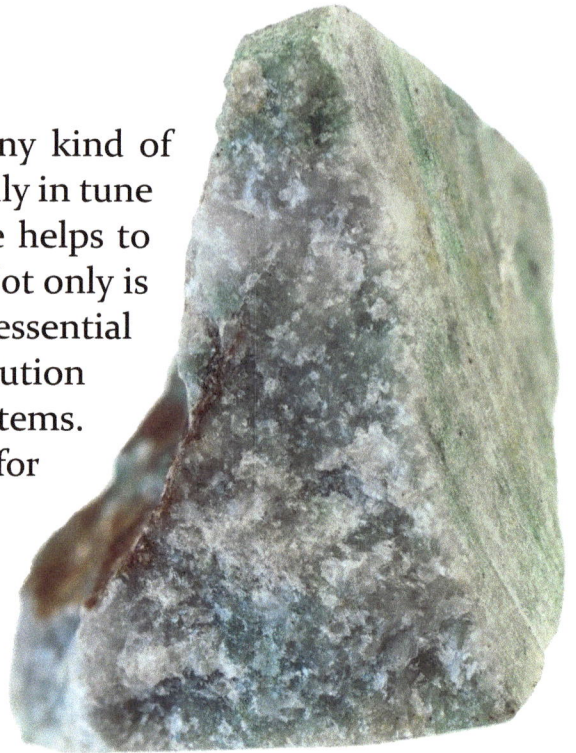

Zodiac Association:	Aries
Common Places Found:	China, India, Brazil, South Africa, Siberia, Italy, & the United States (Delaware)
Chemical Composition:	Silicone dioxide (SiO_2) plus inclusions of chromium; containing muscovite mica
Chakra Association:	Anahata

Properties of Healing

Green Aventurine helps to spark creativity with your Being. It fosters motivation and leadership, allowing

Green Aventurine, Green Quartz, Chrysocolla, Nephrite Jade, Jasper, Amazonite

you to make sound decisions quickly and effectively. As it supports the heart and chest, it is an amazing healer of stress related headaches, poor or stagnant blood, irregular periods, high and low blood pressure, agitation, irritability, insomnia, and imbalanced emotions.[41] [42] Essentially, it is a stabilizer, bringing you into your best rhythm.

Green Aventurine can be a valuable instrument for detoxing the entire body because it energizes and regulates systems. Symptoms surrounding chemotherapy and post-chemo drugs can be alleviated using Green Aventurine. Allergies, acne, psoriasis, eczema, and urinary tract infections can be addressed when using this stone in the form of an elixir[43].

Spiritual

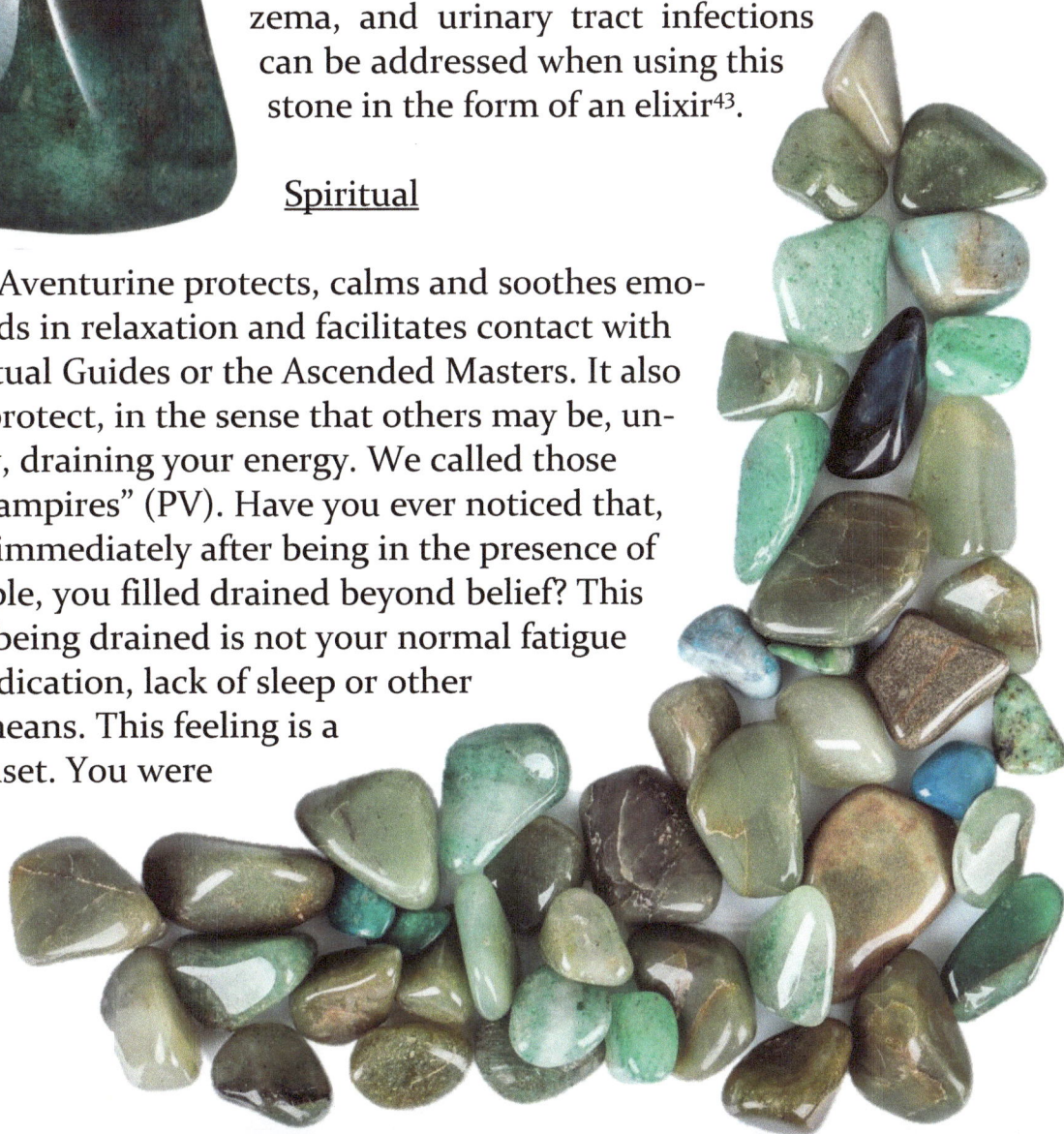

Green Aventurine protects, calms and soothes emotions. It aids in relaxation and facilitates contact with your Spiritual Guides or the Ascended Masters. It also works to protect, in the sense that others may be, unknowingly, draining your energy. We called those "psychic vampires" (PV). Have you ever noticed that, during or immediately after being in the presence of some people, you filled drained beyond belief? This feeling of being drained is not your normal fatigue due to medication, lack of sleep or other physical means. This feeling is a sudden onset. You were feeling great or fairly ok and once a certain person enters the room, you want to sleep, rest your eyes or lie down for a moment.

Consider our young ones. Many guardians tend to associate their after school fatigue with wild accusations such as drug use or intimate activities. It may be accurate in some cases, but I urge you to get even more curious and ask deeper level questions. It is tempting and quite easy to remain focused on surface level inquiries. For guardians, it is sometimes more comfortable, however, stay the course. Inquire about room and body temperature, skin color changes and body energy levels. You may ask, "How did you feel when you walked in the room? How did you feel when you walked out? Does your head ache?" etc. Your young one may be encountering multiple vampires. Since stones may be worn in the form of jewelry or carried in our pockets, there's no harm in allowing them to take the stones to school. Spiritual atonement and balancing works on a more subtle level. Therefore, be observant and take keen-eye notes of your young one's actions following the start of crystal healing activities. Also, ensure that your young one understands and is in agreement with the use of Crystal Therapy. The healing is much more effective when positive energy from the one seeking balance works in conjunction with the stones.

Those who find themselves casualties of PV encounters are generally more 'open' to energy exchange. They are easily affected by another's state of wellness. When someone is sad, whether you know them or not, your senses are affected as they pull on your energy threads. You may even have a difficult time watching horror or suspenseful films due to the excitement portrayed by the characters; sending your systems into energy overload. It may also be common for you to have been labeled "soft" or "in your feelings," more often than not. You are essentially a sensitive person, meaning your chakra portals are more open to Qi, or mana, transfers than others. This should not be viewed as a negative trait. Instead, it is more beneficial to develop ways to strengthen your heart chakra; solidifying habits to maintain your healthy balance. This can be accomplished with Crystal Healing Therapy.

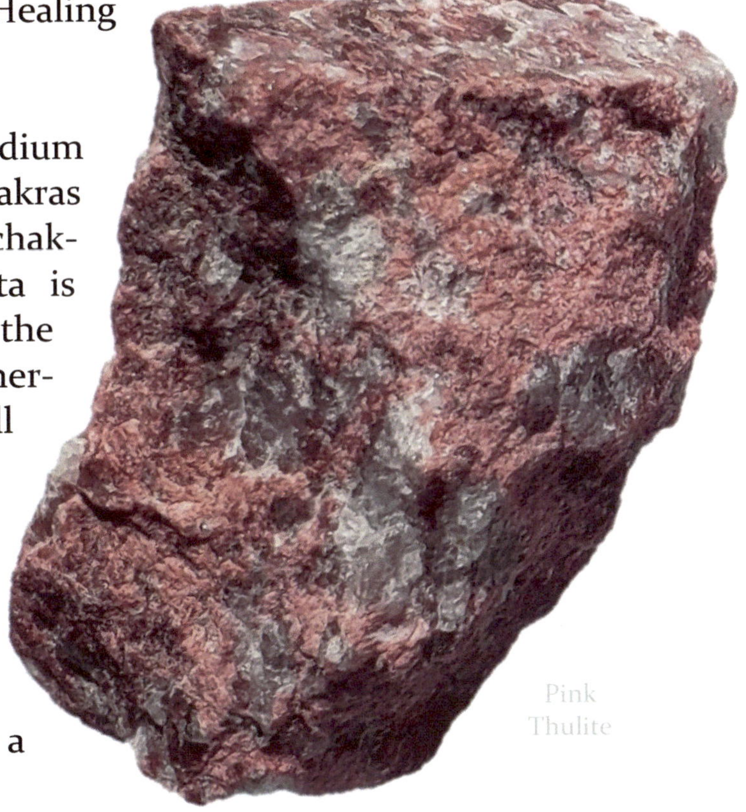
Pink Thulite

Anahata is the medium through which the lower chakras merge with the upper trinity chakras. It is vital that Anahata is balanced, for if it is not, the seamless merge of healthy energies from the base chakras will not be able to communicate successfully with the trinity above. Dis-ease will manifest and imbalances will become apparent. Healing and uniting with Green Aventurine sets the stage for a renewed existence.

When meditating with Green Aventurine, place the stone on Anahata. This point is slightly below the heart muscle, on your sternum, in the center of your chest.

<u>Cleansing and Cautions</u>

Because any size Green Aventurine can hold a large amount of negative energy, using it daily, you would only need to cleanse it once a month. You would position it under running water for an hour. This is necessary to rid your stone of unwanted energies. Subsequently, you may choose to cleanse it once a week, placing it under running water for 15-minutes. You do not have to run the faucet on full blast, it just needs a constant flow. To recharge your stone, just place it on a bed of Hematite. You may also charge in daylight. It is advisable to not leave it in direct sunlight because its color could fade.

If you find that you are having trouble with eliminating waste from the body in forms of sweating or bowel movements, reduce the time you an in contact with your stone. Keep in mind that this is a detoxing stone. It pulls impure energies and minerals from the body. Medication absorption can be negatively affected if you use Green Aventurine within 2-hours after taking prescribed medication or highly concentrated herbal remedies.[44]

Green Tourmaline

Additional Gemstones that can be used to Cleanse & Balance Anahata:

Garnet, Green Tourmaline, Pink Tourmaline, Green Fluorite, Rose Quartz, Rhodochrosite, Thulite, Chrysocolla, Green Opal and Green Jasper.

Citrine

Citrine is a crystal of the Quartz family used to balance and make whole. Citrine's color palate is unique in that it can range from a translucent light gold to a rich, dark brown. Citrine's energy is like that of the Sun; warming, comforting, penetrating, energizing, restorative, and life giving. Citrine's energy specifically attracts to us the bountiful riches of the Earth. Riches come in many forms; from knowledge, to love, monetary, emotional, material, and other types of support.

When balancing and performing healing work using Citrine, one is able to reach a state of consciousness that, even while walking through your daily, you can invoke the Law of Attraction and draw to you the deep desires of your heart. As you are balanced and whole, your desires will be of Light and love. You would not, in a peaceful-secure-grounded state of Being, wish negative or harmful things to come to fruition. If you find that you do, then you must accept that you have more work to do in healing your Body, Mind and Spirit. The Divine and positive energies are attracted to 'like' energies. If you are in a negative state of mind and negative state of being, you are an opposing force to the Light. This causes separation and imbalance. Citrine is one to help bring you back to your center, bridging the connection of Anahata to your lower chakras and creating a protected state of wholeness.[45]

Zodiac Association:	Aries, Gemini, Leo, & Libra
Common Places Found:	Brazil, Himalayas, European Alps, & the United States (Arkansas & Texas)
Chemical Composition:	Silicone dioxide (SiO_2) plus iron. The iron in Citrine affects the lung and spleen Qi (mana/energy). Citrine can also contain traces of sodium, lithium, calcium, aluminum and magnesium.[1]
Chakra Association:	Manipura

Citrine

Properties of Healing

Citrine is nicknamed the "money stone" because of its unrivaled ability to bring the brilliance of life force energy to your Solar and quickly emit your attraction vibrations; drawing wealth and abundance to you. It encourages learning, teaching, studying, awareness, writing, problem solving and new beginnings. If your wealth is meant to come from the stock market, it is likely that you will have to learn or study the subject of trading stocks.

Citrine supports the digestive system by relieving bloating, flatulence, lethargy, and fatigue. It also helps to relieve mental confusion and the inability to concentrate. It can ease the symptoms of those suffering with early onset dementia. Citrine is also used to regenerate and tone the skin. It helps to regulate blood sugar in those with diabetes, heal postsurgical wounds and eases poor circulation associated with Raynaud's syndrome. Other benefits include strengthening a weakened immune system and correcting hypothyroidism. In the areas of reproduction, it assists with a diminished libido and increases fertility in women.[46] [47]

Lemon Opal

Spiritual

Citrine is an amazing stone to partner with to build your self-worth and self-esteem. It promotes confidence and uplifts your spirits. It cleanses you of emotional toxins, such as anger, and balances your Yin and Yang. It works to give a strong sense of hope and encouragement to those who are depressed and disappointed with themselves. It helps to build internal support for those who may feel like failures, receive no acknowledgment and cannot seem to gain a footing.[48]

Meditating to balance Manipura calls for you to place Citrine on your solar. This chakra point is approximately 2-inches above the naval or bellybutton.

Cleansing and Cautions

Cleanse Citrine in running water. Citrine holds more heat than other Quartz Crystals and needs to be cleansed or discharged more frequently. If you are working with Citrine daily, cleanse every other day. Allow it to charge in daylight when not in use. Alternately, you may place it in direct sunlight for 1 to 2-hours; on a sunny day, 1-hour, on a partially cloudy day, 2-hours.[49]

Additional Gemstones that can be used to Cleanse & Balance Manipura:

Lemon Quartz, Lemon Opal, Yellow Jasper, and Golden Tiger's Eye.

Golden Tiger's Eye

Carnelian (Orange)

Carnelian is a variety of Chalcedony in the form of orange, red, pink, or brown pebbles. The orange color of Carnelian is the product of Fire moving into the Earth, thus resulting in an orange hue. Steadily burning, the fire removes the impurities and only leaves behind the radiance. Carnelian conveys the ability to look at one's life with compassion, reflect on one's boundaries and begin to change. It is extremely helpful in overcoming the emotional and mental trauma of abuse; as it grounds and anchors you into the present moment. It resonates with Svadhisthana, the site of creativity and fertility.[50] [51]

Zodiac Association:	Cancer, Leo & Taurus
Common Places Found:	India, Brazil & Uruguay
Chemical Composition:	Silicone dioxide (SiO_2) plus iron.
Chakra Association:	Svadhisthana

Properties of Healing

Carnelian balances the energy flow of Yin and Yang, along with the body's blood flow. It improves absorption of vitamins and nutrients; which is important for the elderly. Carnelian's vibrations help to ward off the negative opinions and influences from those closest to us who tend to focus on illnesses; be they mental or physical. It

encourages individuals who find themselves in unsupported situations to dissipate pessimism and doubt.

Carnelian helps to regulate breathing; treating shortness of breath and diaphragmatic constriction. It addresses acute pain and inflammation, supports kidney functions, menopausal signs and symptoms, premature ejaculation and post-partum concerns. Because it nourishes the blood, Carnelian supports the health of fertility. It accomplishes this by mitigating impotence, poor memory and the inability to release difficulties. When the mind is at peace, creativity in all forms, has spiritually 'grounded', fertile soil from which to mature.[52]

It clears the heat in the blood, thus reducing high blood pressure, frequent bleeding gums, anxiety and stress disorders. Breaking up stagnant energy of the lower region, Carnelian helps to relieve or remove fibroids, ovarian cysts, abdominal pain, lower back pain and kidney stones. Research in the field of Chinese Stone Medicine shows Carnelian can also be used to increase hemoglobin.[53] Effective at healing your liver, Carnelian helps to relieve and prevent gallstones and jaundice.[54]

Spiritual

Carnelian helps you to expel disturbing memories and to let go of the past. It relieves nervousness that accompanies excitement. It levels the energy of individuals who are facing a burnout or who have already "crashed and burned". Carnelian clears your thoughts of negative energy, such as anger, fear sorrow, jealousy, disappointment; your mind is free to connect to the spiritual or Divine Realm.

When meditating to remove blockages and balance Svadhisthana, place the stone on your chakra point. It is about 2-inches below your navel.

Cleansing and Cautions

Hold Carnelian under warm running water to draw out the heat that has accumulated from your healing session. The heat from the warm water provides the Yang that has accumulated to be discharged. Charge by placing it on a Selenite or Quartz Crystal. You may also place it in an Amethyst geode. After washing, you may also place it outside in the daylight for a couple of hours while it is drying, then place on a Selenite, Clear Quartz, or Amethyst Crystal to complete its charging.[55]

Additional Gemstones that can be used to Cleanse & Balance Svadhisthana

Orange Calcite, Orange Quartz, Petrified Wood and Amber

Red Jasper

Red Jasper is a member of the Quartz family whose red components are the result of iron oxide. Muladhara is The Foundation, your Base Chakra. Red Jasper is a stone of nurturing energy. It is a mood normalizer. It promotes cooperation within your Being as well as those you encounter. Red Jasper has the ability to create its own protective barrier around you, allowing your Being to work, mature, heal and find balance while

Zodiac Association:	Aries & Taurus
Common Places Found:	India & Brazil
Chemical Composition:	Silicone dioxide (SiO_2)
Chakra Association:	Muladhara

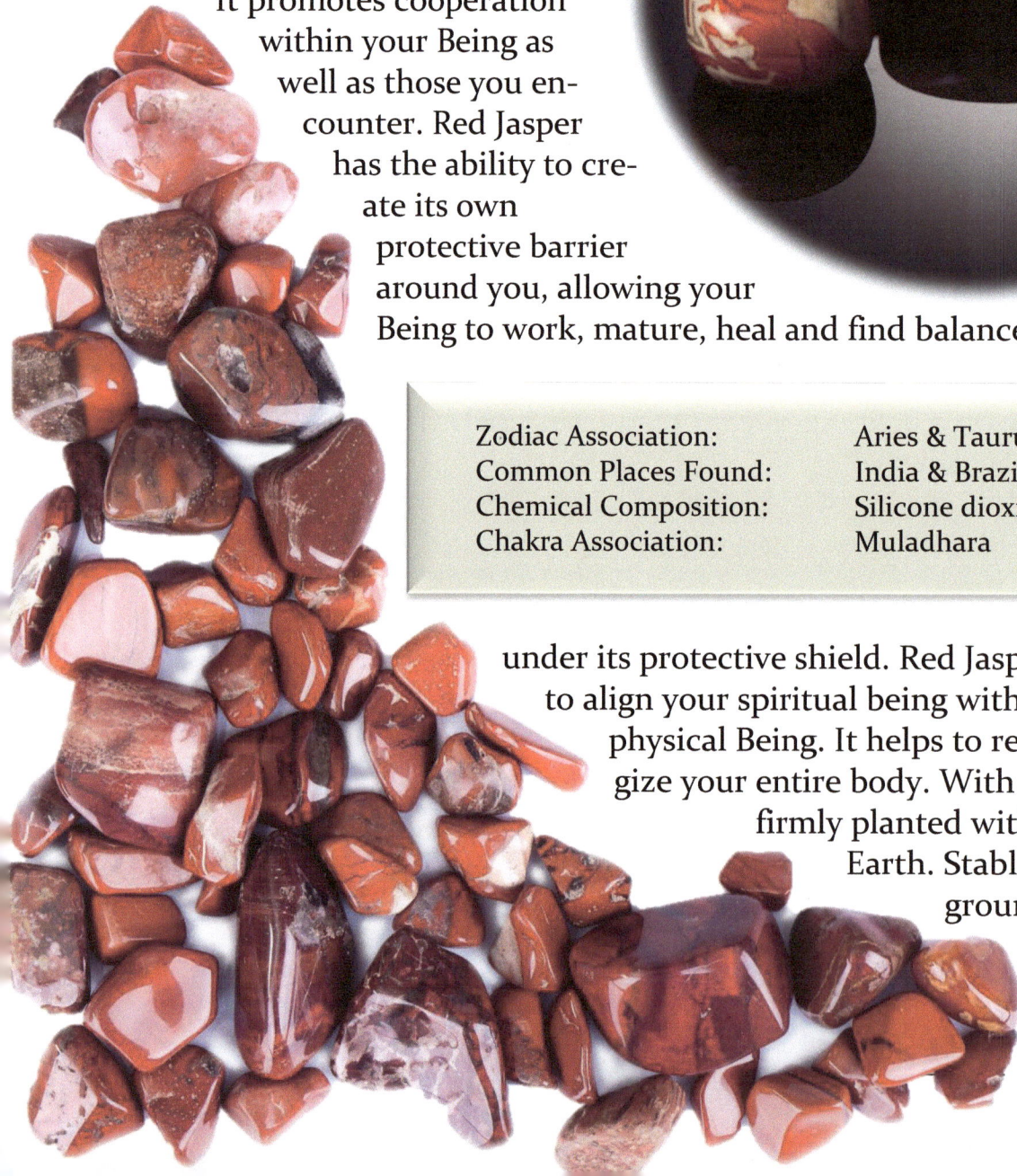

under its protective shield. Red Jasper helps to align your spiritual being with your physical Being. It helps to re-energize your entire body. With your feet firmly planted within the Earth. Stable grounding allows you to build a temple able to withstand the interference of both stormy and sunny weather.

Properties of Healing

Red Jasper breaks apart stagnant blood affecting the heart. This eases the flow of blood, initiating effective nutrition delivery. This also cools the blood. It helps to cure insomnia and reduce high blood pressure, shrink and relieve symptoms associated with hemorrhoids, varicosities and phlebins. Due to its grounding abilities and calm nurturing nature, Red Jasper effectively treats anxiety, mood disorders, irritability and nervousness.[56]

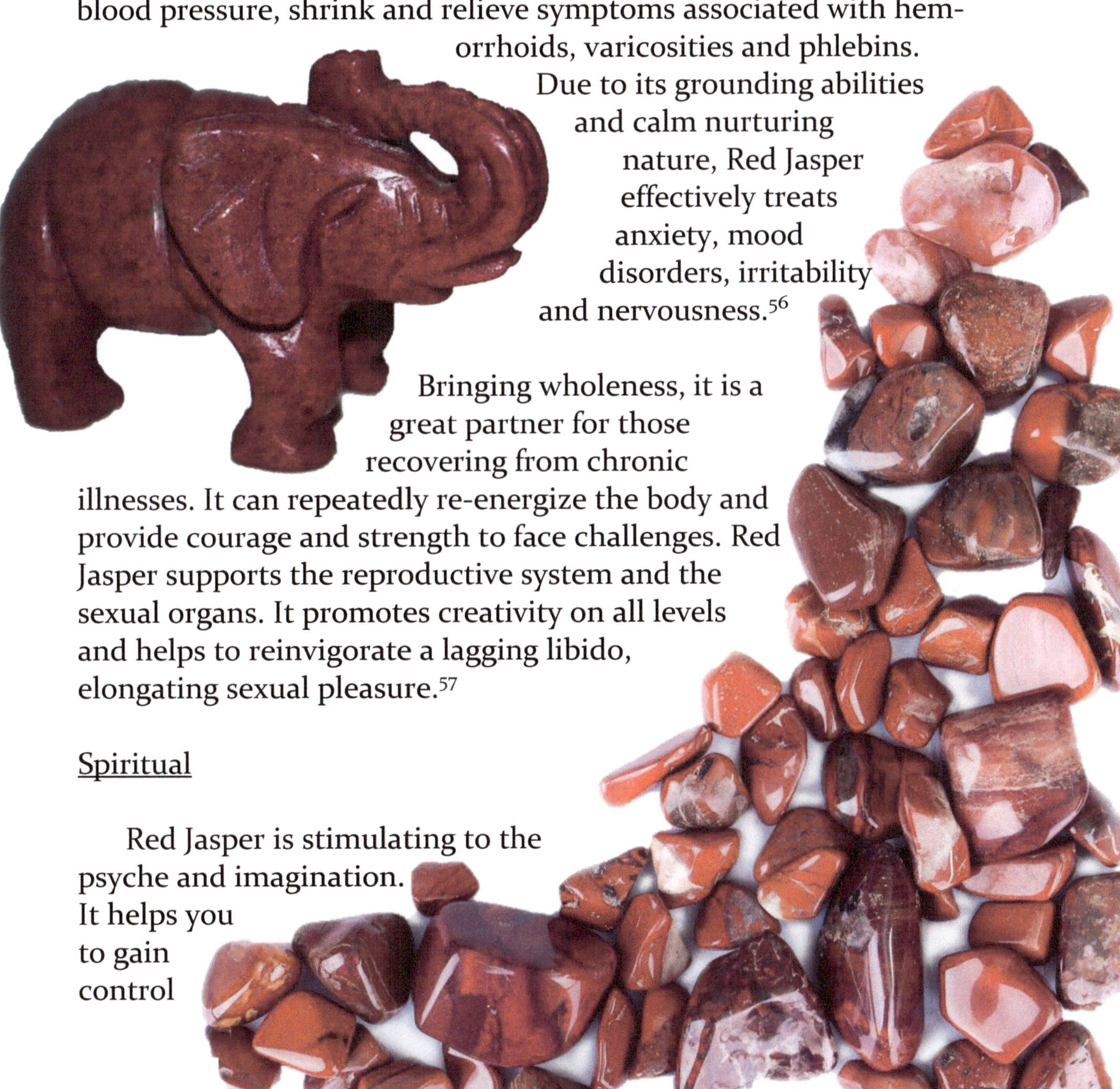

Bringing wholeness, it is a great partner for those recovering from chronic illnesses. It can repeatedly re-energize the body and provide courage and strength to face challenges. Red Jasper supports the reproductive system and the sexual organs. It promotes creativity on all levels and helps to reinvigorate a lagging libido, elongating sexual pleasure.[57]

Spiritual

Red Jasper is stimulating to the psyche and imagination. It helps you to gain control of highly 'spirited' encounters and situations before they get out of hand. As confidence settles in, you become more assertive; you find it easy to face the truth and be honest with yourself. A shamanic stone, it protects the body and boundaries when journeying to other worlds. When working with spiritual matters, carry Red Jasper with you for protection and manifesting more of your authentic self, here, in Earth realm.[58]

When meditating or balancing your chakras, place Red Jasper on Muladhara. This is posterior at the tail bone, at the perineal, or anterior about an inch below the bikini line. It works best with direct skin contact. Because Red Jasper works slowly, meditation work should be longer than 5-minutes at a time.

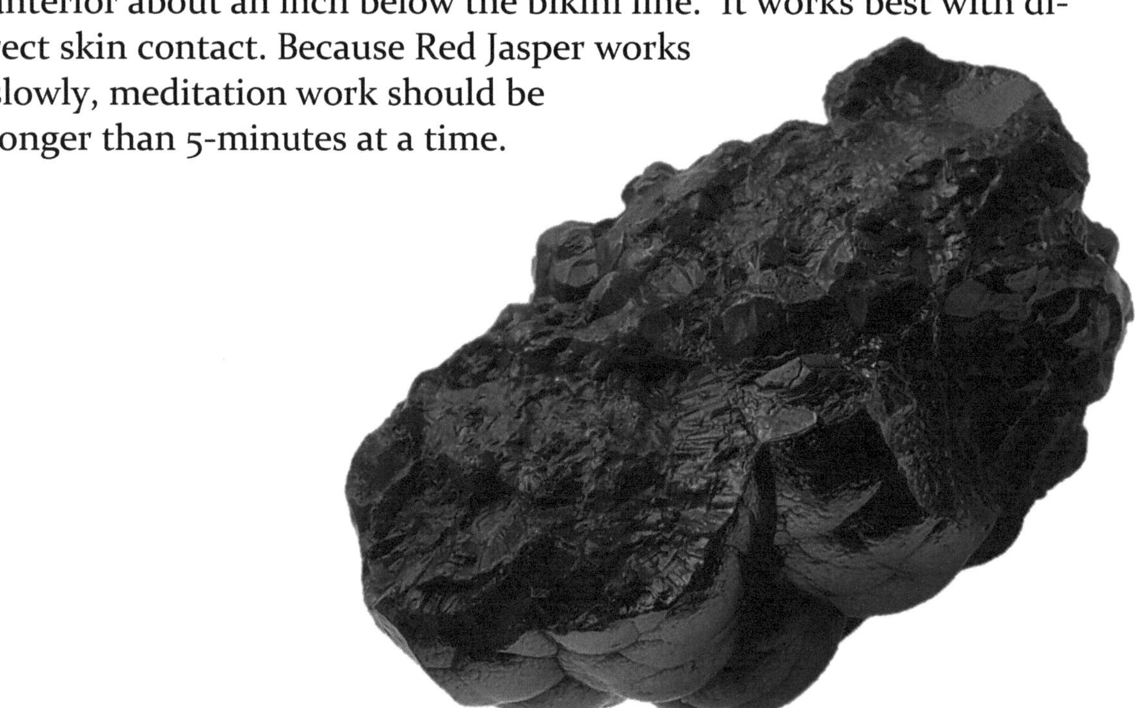

Hematite Botryoidal (Kidney Ore)

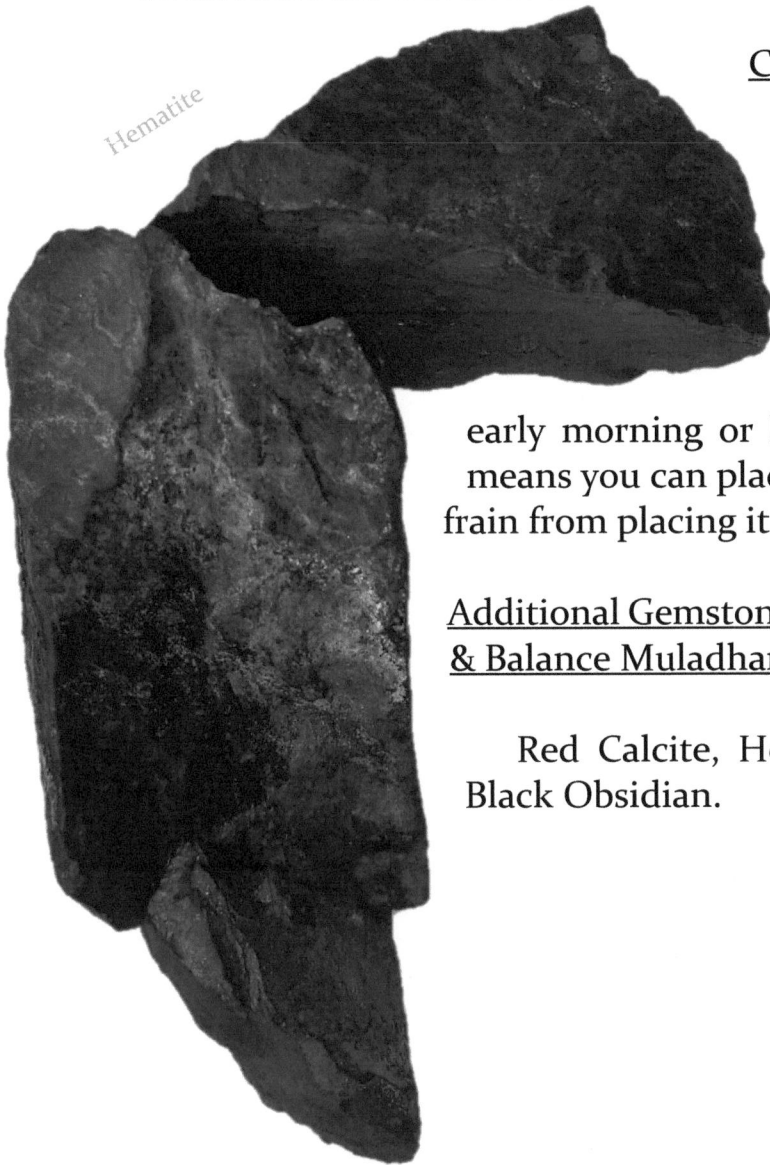

Cleansing and Cautions

The cleansing of Red Jasper is performed once a month by holding it under warm running water. Recharge on a Hematite stone or place it outside in the early morning or late afternoon sunlight. This means you can place it in direct sunlight, but refrain from placing it in the high noon sunlight.

Additional Gemstones that can be used to Cleanse & Balance Muladhara

Red Calcite, Hematite, Smoky Quartz, and Black Obsidian.

Black Tourmaline (For Completion)

Black Tourmaline, also known as Schorl, is added to your meditation healing session for protection and stabilization of energy. This stone completes the circle of healing during your crystal healing therapy session; designed to balance all of your chakras to build a healthy happy balanced Being. A powerful protector, Black Tourmaline can be used in any type of healing therapy. Carried with you, it can prevent a ward of negative events.

When you let the Universe know you intend to carry Black Tourmaline with you at all times, you will start to notice that you miss engagements or you take alternate routes. Incidentally, there may be a sudden onset of traffic or your client cancels. I encourage you to adopt a different perspective. Because you are now calling on the Protective Energy of the Universe, you have now stepped into a plane of existence where certain negative situations will be avoided if they do not serve in your best interest. There are instances where a negative situation serves in your best interest. For example, a trip to the dentist for a root canal, a lecture from your guardian or cleaning up the mess the dog left on the carpet. There is learning and growth opportunities in those "negative" situations.

Think for a moment on the traffic that causes you to take an alternate route. You may find my shop on an alternate route. You may realize that you were unknowingly low on gas and taking that alternate route prevented you from stalling on the highway. These are events that, with a change in perception, can reveal what's possible when you invoke the protection energies of Black Tourmaline on a regular basis.

Zodiac Association:	Libra
Common Places Found:	Brazil & Pakistan
Chemical Composition:	Sodium iron aluminum borosilicate hydroxide ($Na(Fe,Mn)_3Al_6B_3Si_6O_{27}(OH)_3(OH,F)$). Its hydroxides purge impurities. Black Tourmaline can also contain traces of magnesium, potassium, or calcium.
Chakra Association:	All

Properties of Healing

Black Tourmaline works with the kidneys and the lungs to release energy to the exterior. It directs invading energies and toxins away from the body's DNA to give the system the time it needs to repair itself at the cellular level. It moves the energy to expel, not only internal but also, external pathogenic factors to the outside of your protected energy barrier. This means it has the ability to clear environmental toxins, like metals and mercury. It battles autoimmune conditions, shingles, chronic infestations, mental fogginess, and assists those with autism. Black Tourmaline is a barrier breaker, as it removes blockages. It works to heal hepatitis, gastritis, and heart conditions, such as clogged arteries.[59]

For you sessions, Black Tourmaline can be placed at the knees or feet with terminations or points of the stone, facing towards the toes, downward, away from the body. During Crystal Healings, this directs negative energies out and away from the body.

It protects against evil curses by absorbing negative energy in propelling it. You may also create a Black Tourmaline Talisman to work against black magick or magick intended to cause you debilitating harm.

Black Tourmaline can also be used to open and develop Ajna. Reason being, it absorbs Divine Light into the brain, offering the wisdom of "finding yourself"; promoting knowledge of Thy Self.[60] When developing your Third Eye, Black Tourmaline protects against psychic attacks. You are opening your mind and spirit to the astral plane, both channeling and projecting your energy, therefore, protection is vital. It is important to remember that all energy is not good energy.[61]

When meditating for cleansing and balancing your chakras, place Black Tourmaline at the bottom of your feet. With your feet together, place your Black Tourmaline stone about 2-inches away from the body's center line between both feet. Ensure the stone is an extension of your center line and is not perpendicular, crossing your center line. Black Tourmaline's construction can appear to have multiple black strips fused together. Positioning your stone in this manner while meditating is efficient in facilitating the flow of energy. If not perfectly in position, Black Tourmaline will still perform as one needs during their session.

Cleansing and Cautions

To cleanse Black Tourmaline hold it under running water. Recharge on a Quartz stone or Selenite. Do not allow Black Tourmaline to be submerged in water. Because of its porous composition, the stone may disintegrate quicker than by natural means. Cleanse every two weeks or once a month, depending upon the frequency of use. Do not place it in the sun. Black Tourmaline will eventually break down or crumbled, whether or not you cleanse in water. When that breakdown happens or if the stone separates in any way, gather its pieces and bury your stone, for it has finished its job in the physical Earth realm.

Chapter 5: Healing Stone's Purpose

If you find that any of your stones have fallen and split in half or in pieces, just as the procedure with Black Tourmaline, gather up as many as you can and bury them. They have completed their works. You are able to, even while working with the stones, buy or collect additional stones to add to your healing and strengthening sessions. Your Crystal Healing Therapy sessions can be tailored to be as unique as you are.

Clear Quartz Crystal for divine unification, Amethyst for Sahasrara development, Charoite for Ajna clarity (seeing), Blue Kyanite for Vishuddha communication, Green Aventurine for Anahata strengthening, Citrine for Manipura manifesting, Carnelian for Svadhisthana clearing, Red Jasper for Muladhara grounding, and Black Tourmaline for protection ensures your chakras will become balanced and operate in unison. I hope that this text has sparked a question or two and provided answers to others. As a well-balanced Being, you will be able to live the best life possible, during this lifetime, in this amazing Universe.

Endnotes

[1] Webster, pp. 66
[2] Korotkov, 2019
[3] Stein, pp. 10-11
[4] Stein, pp. 9, 12
[5] Rand, pp. 19
[6] Stein, pp. 13
[7] Stein, pp. 14
[8] Webster 2, pp. 6
[9] Webster, pp. 44
[10] Avalon, 1834
[11] Webster, pp.74
[12]Banzhaf, pp. 24
[13] Webster, pp. 73
[14] Wilcox, 2019
[15] Webster, pp.73
[16] Webster, pp.73
[17] Fondin, 2017
[18] Fondin, 2017
[19] Webster, pp.71
[20] Webster, pp.71
[21] Webster, pp.71
[22] Webster, pp.70
[23] Webster, pp.69
[24] Halevi, pp.25
[25] Webster, pp.69
[26] Guhr, pp. 10
[27] Guhr, pp. 13
[28] Guhr, pp. 15
[29] Guhr, pp. 17
[30] Perrakis, pp. 101
[31] Guhr, pp. 25
[32] Guhr, pp. 27
[33] Schumann, pp. 31
[34] Natterer, pp. 245-502
[35] Raphaell, pp. 49-50
[36] Ahsian, pp. 224-225
[37] Melody, 363-364
[38] Franks, pp. 325-326
[39] Franks, pp. 326
[40] Hall, pp. 90
[41] Permutt, pp.48
[42] Franks, pp.284
[43] Franks, pp. 284
[44] Franks, pp.284
[45] Raphaell, pp.86
[46] Permutt, pp. 41
[47] Franks, pp. 156
[48] Permutt, pp. 41
[49] Franks, pp. 157
[50]Franks, pp. 196
[51] Hall, pp. 82
[52] Franks, pp. 197
[53] Franks, pp. 197
[54] Franks, pp. 198
[55] Franks, pp. 199
[56] Franks, pp. 189-190
[57] Hall, pp. 79
[58] Hall, pp. 81
[59] Franks, pp. 270
[60] Raphaell, pp. 132
[61] Franks, pp. 270

Petrified Wood

Your Personal Notes

Your Personal Notes

References

[Ahsian, pp.] Ahsian, Naisha and Simmons, Robert, *The Book of Stones,* (Berkley, CA: North Atlantic Books, 2007).

[Avalon, 1834] Avalon, Arthur aka Sir John Woodroffe, *The Serpent Power: The Secrets of Tantric and Shakta Yoga,* (Create Space Independent Publishing Platform, 1834).

[Banzhaf, pp.] Akron & Hajo Banzhaf, *The Crowley Tarot: The Handbook to the Cards,* (Stanford, CT: U.S. Games Systems, Inc., 2014).

[Franks, pp.] Franks, Leslie, *Stone Medicine, A Chinese Medical Guide to Healing with Gems and Minerals,* (Rochester, Vermont, Healing Arts Press, 2016).

[Freshwater, 2017] Freshwater, Shawna, Ph.D., *1st Chakra Root Muladhara,* (Spacious Therapy, Nov. 20 2017). [Accessed: June 20, 2019] <https://spacioustherapy.com/1st-chakra-root-muladhara/>

[Fondin, 2017] Fondin, Michelle, *Your Inner Truth with the Fifth Chakra,* (The Chakra Center, 2017).

[Guhr, pp.] Guhr, Andreas & Nagler, Jörg, *Crystal Power: Mythology and History,* (Findhorn, Great Britain: Earthdancer Books, 2006).

[Hall, pp.] Hall, Judy, *Crystal Healing,* (London, Octopus Publishing Group Ltd., 2005).

[Halevi, pp.] Halevi, Z'ev ben Shimon, *Kabbalah and Astrology,* (Kabbalah Society, 2009).

[Korotkov, 2019] Korotkov, Konstantin, *Curriculum Vitae Dr. Konstantin G. Korotkov,* (Louisville, CO: Bio-Well, 2019). [Accessed: June 20, 2019] <https://www.bio-well.com/gb/company/about_korotkov.html>

[Melody, pp.] Melody, *Love is in the Earth,* (Wheat Ridge, CO: Earth-Love Publishing House, 1995).

[Natterer, pp.] Natterer, Fabian D. and Zhao, Yue and Wyrick, Jonathan and Chan, Yang-Hao and Ruan, Wen-Ying and Chou, Mei-Yin and Watanabe, Kenji and Taniguchi, Takashi and Zhitenev, Nikolai B. and Stroscio, Joseph A, *Strong*

Asymmetric Charge Carrier Dependence in Inelastic Electron Tunneling Spectroscopy of Graphene Phonons, (Phys. Rev. Lett., Volume 114, Issue 24, pp. 245502, June 16 2015). DOI: 10.1103/PhysRevLett.114.245502, URL: https://link.aps.org/doi/10.1103/PhysRevLett.114.245502

[Paulson, 1991] Genevieve Lewis Paulson, *Kundalini and the Chakras*, (Llewellyn, 1991).

[Perrakis, pp.] Perrakis, Athena, Ph.D., *Crystal Lore, Legends & Myths,* (Beverly, MD: Fair Winds Press, 2019).

[Rand, pp.] Rand, William Lee, *Reiki The Healing Touch First and Second Degree Manual,* (Southfield, MI: International Center for Reiki Training, 1991).

[Raphaell, pp.] Raphaell, Katrina, *Crystal Enlightenment: The Transforming Properties of Crystals and Healing Stones,* (Santa Fe, NM: Aurora Press, Inc., 1985).

[Schumann, pp.] Schumann, Walter, *Gemstones of the World*, (New York: Sterling, 1976).

[Stein, pp.] Stein, Diane, *Essential Reiki: A Complete Guide to an Ancient Healing Art*, (Berkeley, CA: The Crossing Press, 1995).

[Webster, pp.] Webster, Richard, *Aura Reading For Beginners: Develop Your Psychic Awareness for Health & Success*, (Woodbury, Minnesota: Llewellyn Publications, 1998).

[Webster 2, pp.] Webster, Richard, *Pendulum Magic for Beginners*, (Woodbury, Minnesota: Llewellyn Publications, 2015).

[Wilcox, 2019] Wilcox, Stephanie, *How to Open your Third Eye*, (Psychic Guru, 2019). [Accessed: June 20, 2019] <www.psychicgurus.org/how-to-open-your-third-eye>.

Illustrations & Photos:

(Unless otherwise stated, all photos are the property and works of Liam J. Adair)

iStock Photos: [pp]
Pages: v,viii,2,3,5,7-0,13,14,19-21,23,26,29,30,32,34-39,46,48,53-54,57,64-66,74-76,83-87,90-93,95

Pixabay Photos: [Artist,pp];
Activedia,16; KatinkavomWolfenmond,79; Alusruvi,62; annca,ii; qimono,47&51; geralt,47&49; Hermann,45; istones,67; malachit-obchod,xi&42; marsjo,1&42; mohamed_hassan,4; skeeze,i; SuzyT,52; TBIT,53; uhi_noko812,61; Varga,11&42; WebLab24_Siti_Web,87.

Shutter Stock Photos: [pp]
Pages: 15,25,27-28,33,40,55,67-69,71-73,75-76,79,80

Index

About the Author Liam J. Adair

Liam launched his holistic health career in 2019 and built a successful private practice, Wholesome Healing, focused on an individual's physical, emotional, mental, and spiritual well-being. As the first Black Transgender Board Certified Holistic Health Practitioner in the State of Texas, as well as a graduate of Quantum University, he excels at applying principles of quantum medicine and energy balancing mindsets to the life coaching field.

In 2020, he joined his local and national LGBT Chamber of Commerce and began focusing on working with corporations to increase employees' motivation, engagement, and fulfilment in both their home and work environments.

As of today, he is a well-established author of "Fear Rules My Life" and "Crystal Healing for Your Chakras: The TRUE Call of Nature"; published to give readers the ability to begin and maintain their own wellness journey. Having created two online wellness coaching courses, "The Art of Resilience: How to Conquer Challenges" and "Walk in Your Truth: Unlocking the Authentic You" Liam consistently pushes his own perceived limitations, surpassing all expectations. Encouraging his clients to do the same, he has worked with many individuals throughout the globe; from the United States to Africa, India, Europe, South Korea, Japan and Canada.

Citrine Spirit Cactus

Red Garnet

Shungite

Quartz Generator
(crafted)

www.ingramcontent.com/pod-product-compliance
Lightning Source LLC
Chambersburg PA
CBHW040735150426
42811CB00063B/1636
* 9 7 8 0 5 7 8 6 2 0 9 8 5 *